# RECONSTRUCTION AFTER DISASTER

# Reconstruction After Disaster

Issues and Practices



*Edited by*

ADENRELE AWOTONA

*University of Newcastle upon Tyne*

**Ashgate**

Aldershot · Brookfield USA · Hong Kong · Singapore · Sydney

Published by
Ashgate Publishing Ltd
Gower House
Croft Road
Aldershot
Hants GU11 3HR
England

Ashgate Publishing Company
Old Post Road
Brookfield
Vermont 05036
USA



**British Library Cataloguing in Publication Data**

Reconsruction after disaster : issues and practices
1. Disaster relief 2. Emergency management
I. Awotona, Adenrele
363.3 ' 48

**Library of Congress Catalog Card Number:** 96-80381

ISBN 1 85972 551 1

Printed and bound by Athenaeum Press, Ltd.,
Gateshead, Tyne & Wear.

# Contents

# List of figures

# List of tables

# Notes on contributors

**Adenrele Awotona** is a consultant in the field of the Built Environment (architecture, urban design, and human settlements development planning). He is currently the Director of the Centre for Architectural Research and Development Overseas (CARDO) and the Director of Postgraduate Studies in the Department of Architecture at the University of Newcastle upon Tyne, England. He was formerly the Head of the Department of Architecture at the University of Jos in Nigeria, Honorary Consultant Architect to the National Universities Commission of Nigeria, the Head of the Department of Urban Planning and Architecture at the University of Garyounis in Benghazi, Libya and the Director of the *Architecture and International Development* Programme in the Department of Architecture at the University of Huddersfield in England. He has also been (and still is) a Consultant on projects funded by national and international agencies including the United Nations Centre for Human Settlements (UNCHS), the United Nations Development Programme (UNDP), the Overseas Development Administration (ODA) of the British Government, the British Council and the European Union in Brussels. He has recently been in Windsor Castle (UK) for Consultation on Shelter as part of a sustainable future.

He has practised, researched and taught various subjects in the field of the Built Environment (especially architecture, urban design, urban development planning and housing) in the United Kingdom and Africa (north and south of the Sahara Desert). He has also been on research assignments to North America (United States of America and Canada), South America (Guyana), Europe (Greece, The Netherlands, Sweden and Italy), The Middle East (The United Arab Emirates), Asia (India, Turkey, Singapore and the People's Republic of China) and, the South Pacific (Papua New Guinea). Similarly, he has been invited to advise on curriculum development in University Faculties concerned with the Built Environment internationally. Furthermore, he has been an External examiner for

231 higher degree dissertations and theses (MA, MSc, PhD and DPhil) in several universities world-wide.

Dr Awotona has published 2 edited books and over 120 research papers in books and international journals on various aspects of the Built Environment and he frequently participates in international conferences and workshops. He is a referee for a number of international academic journals and is at present a Member of the International Advisory Board of the Third World Planning Review (Liverpool University Press, UK). Similarly, he has been listed in a number of International Who's Who in Education.

**Ngai Weng Chan** is currently Associate Professor in the Geography Section, Universiti Sains Malaysia, Penang. He has written 5 books, 2 chapters in books, 23 refereed articles in professional journals (another 7 currently being refereed), 4 newspaper articles, 12 unpublished research reports, 1 translation and has completed 26 research projects since 1981. He is actively involved in research dealing with environmental hazards (especially floods), hydrology, sustainability and the effects of humans on climate, as well as currently being involved in several consultancies. Dr Chan is also a Visiting Academic at the School of Geography & Environmental Management, University of Middlesex, UK, where he obtained his PhD (Hazards Management) in 1995. Since July 1995 he has been a committee member of the Malaysian Nature Society (Penang Branch) and is a member of the Institute of British Geographers, the British Hydrological Society, the Royal Geographical Society and the Natural Hazards Society.

**Diane Diacon** trained as a town planner at Nottingham University and qualified in legal and accounting practise, she currently works as Research Officer at the Building and Social Housing Foundation. She has an extensive knowledge of housing in both the United Kingdom and other countries and her publications include *Residential Housing and Nuclear Attack, Homes Above All* (Ed.), *The House that Grows* (Ed.), *Deterioration of Public Sector Housing Stock in the United Kingdom*, *Cities of the Future*, *Innovative Housing Solutions in Beijing*, *Alternative Futures - the reallocation of power* and *Responsibility And The Reallocation of Resources To Meet Global Shelter Needs*.

**Souheil El-Masri** was born in Beirut and graduated from the Faculty of Architecture of Beirut Arab University in 1981. After graduation, he practiced architecture for five years during which he designed several projects in Lebanon and the Middle East. He holds MPhil (1988) and PhD (1992) degrees from the University of Newcastle upon Tyne, England. Both theses focus on Post-war Settlement Reconstruction with special reference to Lebanon. After completing his PhD, Dr El-Masri was granted one-year status as a Guest Member of Staff at the Centre for Architectural Research and Development Overseas (CARDO), University of Newcastle. During this period, he initiated a network and research activities in the field of disaster and human settlements. He was also the editor, with Dr Adenrele Awotona, of a newsletter on Reconstruction After Disaster. In 1994, Dr El-Masri joined the Department of Civil and Architectural Engineering at the University of Bahrain, where he is now Assistant Professor. He has delivered papers at many international conferences and published extensively in the area of disaster and human settlements.

**Michael A. Evans** joined Cranfield Disaster Preparedness Centre (CDPC) in late 1994 after serving in the British Army. He has been involved in the UNDP's disaster management training programme in Eritrea, in disaster relief planning and course preparation in the Caribbean and has undertaken logistics field work in Uganda. He lectures in logistic systems at Cranfield University and at the Bangladesh Defence Services Staff College on relief planning and development. More recently, he has designed and delivered disaster management courses in Jordan and Zambia. His other work areas in CDPC include complex emergencies and the roles of military forces in humanitarian operations and wider peacekeeping. During his military career, he served in Great Britain, Germany, Sudan, Cyprus and Northern Ireland. He has been an instructor at RMA Sandhurst and at the Army Staff College. He commanded an armoured engineer regiment in Germany from where he moved to Headquarters 3rd Armoured Division. His last appointment in the Army was in the Ministry of Defence, dealing with future concepts.

**Nabiul Islam** is currently a Research Fellow at the Bangladesh Institute of Development Studies. His research activites include policy oriented research, of an empirical and theoretical nature, in the following fields: Floods, famines and mass starvation; rural industries; technology and employment; evaluation of rural development projects; base-line survey of flood control projects; rapid rural appraisals of flood control projects and full-impact evaluatoin of flood control projects. He is currently pursuing a PhD at the Flood Hazard Research Centre, Middlesex University, UK (expected submission date November 1996) and has a number of publications including three co-authored books.

**Mulbah G. Johnson** has an MA degree in Housing Studies from the University of Newcastle upon Tyne, UK and is currently the Director of Don Bosco Rehabilitation Programme in Monrovia, Liberia.

**Vladimir B. Ladinski** was born in Skopje, Republic of Macedonia and educated at the Universities of Skopje (Architecture: DipIngArch, 1980), Ljubljana and Anglia Polytechnic University (Urban and Regional Planning: MSc, 1994). He is a Registered Architect in the Republic of Macedonia (1984) and in the UK (1995) and is a member of the Architects and Surveyors Institute, UK (1995). He is currently reading for a PhD on Urban Planning and Design in Earthquake Prone Areas at the Risk and Disaster Management Section, Institute of Earthquake Engineering and Engineering Seismology at the University St Cyril and

Methodius, Skopje, under the supervision of Prof Zoran Milutinovic. Mr Ladinski was an architect in Macedoniaproekt, Skopje between 1981 and 1990 and with Nicholas Ray Associates, Cambridge between 1990 and 1993 and since 1995. His professional experience is in urban planning and design, and architectural design in earthquake prone areas, as well as in refurbishment of historic buildings. His research interests are in post earthquake reconstruction, debris management and recycling in seismic areas with special interest in site vulnerability analysis and development control standards in earthquake prone urban areas.

**James Lewis** has been engaged in research and implementation for disaster reduction since 1970. Numerous publications on environment, hazards, human settlements and development have accompanied and have ensued from the undertaking of overseas assignments in island states of the South Pacific and the Caribbean, and in North Africa, to do with post-earthquake and tropical cyclone reconstruction planning. cyclone resistant construction and climate change and sea-level rise. He is currently engaged upon a series of missions in Bangladesh to do with development planning for disaster reduction and dual-purpose cyclone shelter design and construction and their rural integration. He is also currently involved as architect on the CEC/World Bank/Government of Bangladesh Cyclone Shelter Project. His chapter is based upon his personal views.

**Dennis Parker** is the Head of the School of Geography and Environmental Management at Middlesex University. His general research area in the management of natural hazards and resources, including: flood hazard managment; natural hazards; disaster management; disaster vulnerability of megacities; water institutions and policies; and coastal managment. Recent research projects include: evaluations of flood warning dissemination systems; development control in flood plains; and hazards in the London megacity. He has been and still is Director of Studies for a number of degrees by research as well as teaching courses within the School

Professor Parker has been a member of The International Geographical Union Disaster Vulnerability of Megacities Study Group since 1993, a Fellow of The Royal Geographical since 1985 and a Full Member of the Institute of British Geographers since 1976.

He has fostered a number of overseas research links and regularly acts as a consultant to agencies ranging from an international to local scale. He has published extensively and regularly presents lectures and papers at seminars.

**David Sanderson** works as Project Manager at the Oxford Centre for Disaster Studies. He studied Architecture at Polytechnic in Leicester and Oxford and gained an MSc in Development Practices in 1993. He previously worked as a Researcher at the Centre for Development and Emergency Planning (CENDEP) at Oxford Brookes University. He has worked on a voluntary basis for NGOs on shelter building programmes in Mexico and Colombia. As well as the Peru programme he is currently involved in a disaster preparedness training programme in southern and western Africa, and in a research project involving a comparative study of hazard early warning systems. He was principle researcher for the recent publication *Disaster Mitigation, Preparedness and Response; an Audit of UK Asset* (IT Publications, 1995) and co-author of *Developing Building for Safety Programmes* (IT Publications, 1995).

**Graham Tipple** is Reader in Housing Policy and Development at the Centre for Architectural Research and Development Overseas, University of Newcastle upon Tyne. He graduated with BA and MA in Town and Regional Planning at the University of Sheffield, and PhD at Newcastle University. He is a chartered Town Planner with experience in local government in Zambia and UK. He worked as a lecturer at the Department of Planning at the University of Science and Technology, Kumasi, Ghana, for four years before joining CARDO in 1982. He is the author of more than forty academic journal papers and book chapters and recently wrote the UNCHS (Habitat)/ILO publication, *Shelter provision and employment generation* (1995). He is co-editor of *Housing the poor in the developing world: methods of analysis, case studies and policy,* London: Routledge (1991). He has travelled widely and led research projects in Bangladesh, Egypt, Ghana, India and Zimbabwe.

# Preface

*Adenrele Awotona*

Every year most parts of the world are inflicted with one type of disaster or another. Indeed, images of horror and destruction, dislocation and starvation, as well as those of dying children and grieving women have become common in the newspapers and on television screens. Ironically, in the case of developing countries, the tragic impacts of disasters are further exacerbated by increased level of poverty, rapid and uncontrolled urbanisation, and the continuous changes in the political and economic circumstances. So what can be done to alleviate the suffering of the victims and to support them in rebuilding their lives and homes?

The Television Trust for the Environment (UK) has noted that 'over the past twenty years four million people world-wide have been killed in drought, flood, earthquakes and hurricanes, and close to half the population of the planet have suffered some form of disruption to their lives'. Hence, the United Nations International Decade for Natural Disaster Reduction (1990-2000) has called for concerted local actions for disaster reduction and transfer of scientific and technical knowledge to the people for the purpose.

Similarly, armed conflicts, between and within nations, have since World War II brought about untold human sufferings and massive destruction to millions of people world-wide. Although accurate figures are difficult to come by, Walker (1995, p.68) points out that the International Federation of Red Cross and the Red Crescent Society's assistance, costing some 23 billion Swiss francs (ECU 270 million), is currently going to some 19.4 million people world-wide in international relief operations; 6.6 million of these are refugees or displaced people. This figure excludes the amount being spent annually by the United Nations on peace keeping and relief efforts. What has become alarming is not so much the number of victims and the amount of financial resources involved in disaster relief, but mainly the continuous failure of relief operations to achieve a complete recovery.

Thus disasters, whether natural or man-made, 'cataclysmic' (fast-impact such as earthquakes, floods, cyclones) or 'continuing' (long-term as in the case of wars, droughts and famines), inflict dramatic impacts in terms of loss of life, disruption to livelihoods and destruction to the built environment. This is specially so in the case of many developing countries, where the impacts of disaster are intensified by the lack of adequate relief and emergency plans, lack of resources and by inappropriate reconstruction programmes. So, how can we intervene to mitigate the impact of disaster? How could future disaster be prevented? How could the tragedy turn into an opportunity for development?

*The structure of the book*

This book tackles these key questions by providing an understanding of the different phases of disaster management and planning including emergency, public education and preparedness, mitigation and reconstruction. Amongst the themes which contributors have explored are: concepts and theoretical issues; evaluation of current policies and practice; relief and humanitarian issues; psychological impacts after disaster; tools and techniques for disaster assessment; community-based reconstruction; technological and construction methods; socio-economic reconstruction; conflict and divided city; the culture of war with emphasis on children; and, displacement, resettlement and refugee camps. The emphasis is on understanding the complexity of the subject by identifying a framework in which the important issues of each phase are highlighted and their interaction considered.

The book is based on the papers presented by eminent specialists in these subject areas at the International Seminar on *Reconstruction after disaster* which was organised by the British Council in 1996 in collaboration with the Centre for Architectural Research and Development Overseas, Department of Architecture at the University of Newcastle upon Tyne, UK.

The book consists of nine chapters. In Chapter 1, El-Masri and Tipple examine the concept of natural disasters and their global patterns. They also investigate the linkages between urbanisation and poverty and their effects on human settlements' vulnerability. They advocate a wide range of measures for mitigating natural disasters such as technological, physical, institutional and socio-economic improvements. Chapter 2, by Evans, discusses the process of planning for emergencies. The discussion highlights important issues such as problem definition and task identification; coordination and timing; resource allocation and mobilisation; and, control and communication systems. Chapter 3, by Parker, Islam and Chan, looks at various aspects of flood disaster - their reduction in the context of sustainable development; the principal types of vulnerability; the typical range of flood impacts in developing countries; flood protection and residual hazard; range of choice and vulnerability; and examples of ways of reducing vulnerability and increasing resilience to future flood disasters. They emphasise the importance of assessment of flood impacts as a pre-requisite for effective planning to mitigate flood disasters based on a case study from Bangladesh. In Chapter 4, Lewis scrutinises the 35 projects involving 24 funding agencies for the provision of cyclone shelters in the coastal 'high risk area' of Bangladesh. Amongst the issues covered are the nature of cyclone shelter construction, building maintenance and construction costs. The chapter concludes by recommending types of development to be avoided or in need of detailed scrutiny for vulnerability reduction. In Chapter 5, El-Masri reports the methodological approach employed in the extensive fieldwork which he conducted in one war-damaged village in Lebanon in 1990. The aim of the case study approach was to gain 'insights into local conditions and the real needs of the victims'. He explains the interrelation of the methods used and how they complement each other in order to provide a clear understanding of displacement and reconstruction in rural Lebanon. The chapter also briefly reports the main findings of the fieldwork. In Chapter 6, Ladinski evaluates the long term effects of the reconstruction of Skopje following the 1963 earthquake which still 'remains one of the most devastating European earthquakes of the 20th century, and one which attracted almost unprecedented levels of international co-operation and assistance'. He clearly illustrates how reconstruction activities led to undesirable outcomes and to increased levels of vulnerability by analysing four post-earthquake housing estates. After briefly looking at the housing conditions in Liberia before the war, the effects of the civil war on the various aspects of the human settlements development in that country are then investigated by Awotona and Johnson in Chapter 7. The socio-economic problems resulting from the crisis are also analysed in the wider context of the West African sub-region. The chapter concludes by putting forward various recommendations for developing a post-war national housing and resettlement plan. The provision of typhoon resistant housing (Core Shelter Programme) for the inhabitants of the Philippines Archipelago, 'one of the most disaster prone areas of the world', is the central theme of Chapter 8 by Diacon. Amongst the aspects of the Project examined are its origin, design and construction, modular upgrading, community organisation and participation in the project, and financial issues. The Core Shelter Project in the Philippines provides an excellent example of how an innovative design, efficient organisation and quality control can successfully, and at low cost, mitigate the effect of typhoons. The project benefits the poorest of the poor, whose house and livelihood are regularly swept away by typhoons. The final chapter by Sanderson, reports a project that was carried out in Lima, Peru in South America, the aim of which was to 'identify hazards, vulnerabilities and capacities, with a view to ultimately identifying ways of reducing vulnerabilities and increasing capacities'. The project reinforces the importance of community participation and of strengthening links among the different people involved.

In summary, all the chapters in this book together illustrate that reconstruction after disaster is a heterogeneous process for a variety of inter-related reasons. The first of these is that the built environment is the end-product of a long chain of interactions such as socio-cultural, economic, technological, environmental and administrative aspects. The second is that the complexity of reconstruction is amplified by the urgency and confusion of the situation after disaster, the need for the mobilisation of huge amount of resources (human and material), and the political and international dimensions concerning the issues of aid and relief. The third deals with the requirements for a comprehensive approach in order to identify the needs of the victims and to provide a perspective of how professionals and people can co-operate in order to plan for meaningful reconstruction. Consequently, reconstruction after disaster requires a holistic and evolutionary approach to deal with the variety of aspects (physical and non-physical) of the damaged settlement. The complexity can be tackled when the route of reconstruction filters through these different aspects and their reciprocal relations. This will be achieved by a careful examination of the *pre* and *pro* disaster situation of the affected settlement in order to identify needs, opportunities, difficulties and constraints. In this sense, the approach should make it possible to capitalise on the victims' participation for meeting their own needs, to rectify existing problems in improving conditions and to provide opportunities for enhancing the well-being of the affected community. It is a development process in which adequate input into the built environment could help in economic regeneration, in providing opportunities for co-operation and integration within the affected community, in promoting socio-cultural identity, in providing a link between past, present and future and for establishing a sense of belonging to the built environment.

*Acknowledgement*

I would like to express my thanks to Mrs Julie Elliott, my research secretary, for her assistance in the preparation of this book.

# 1 Urbanisation, poverty and natural disasters:

## Vulnerability of settlements in developing countries

*Souheil El-Masri and Graham Tipple*

### Introduction

In many developing countries, rapid urbanisation and poverty force people to live on dangerous, marginal sites and in poorly built houses. Overcrowding, subdivision and other conversions of dwellings, lack of amenities and deterioration of buildings, commonly characterise such settlements. Urbanisation and poverty in these countries not only increase the spontaneous growth of squatter settlements but also create human settlements which are more vulnerable to natural hazards.

The aim of this chapter is twofold. The first is to clarify the concept of natural disasters and to discuss their global patterns. The second is to analyse the relationships and interconnections among urbanisation, poverty and vulnerability to natural disaster. The analysis will highlight the underlying reasons for fragile, ad-hoc urban growth which, in its turn, increases vulnerability to natural disasters. Understanding conceptual issues and highlighting linkages between disasters and root causes of vulnerability are essential for mitigating natural disasters in developing countries.

### Natural disasters and human settlements

Natural hazards, as part of our environment, can occur anywhere. Earthquakes, floods, volcanoes and violent weather variations, as well as other extreme natural events, can trigger disasters when they interact with vulnerable conditions. Loss of life, disruption of socio-economic activities, and damage to property and natural resources are some of the common outcomes of these events. While natural disasters do not discriminate among people or countries, they affect some parts of the world more severely than others. This is especially true in developing countries, where the effects of natural disaster are exacerbated by vulnerable conditions such as population growth, increased levels of poverty, inadequate construction, and rapid and uncontrolled urbanisation. These conditions make mitigation of natural disasters a complex process requiring comprehensive planning. However, this complexity is exaggerated in many developing countries by the sheer numbers of vulnerable people, stagnant economic conditions, and inadequate resources and institutional frameworks.

## Natural disaster: conceptual issues

Although considerable effort has been devoted to conceptualising the term 'disaster', many divergent definitions appear in the literature. Disaster is perceived differently by social scientists, politicians, environmentalists, relief workers, or journalists. Reasons for the lack of a commonly accepted definition vary and relate to the complexity of natural events. First, natural disasters are the outcome of an interaction between natural events and human actions; in other words, not all natural disasters are exclusively natural. Therefore the conventional classification of disaster, natural (Acts of God) and man-made (Acts of Man), has now been discarded (Wijkman and Timberlake, 1984). Second, knowledge has expanded with changing ideas about causes and effects - direct and indirect. Third, disasters are complex events with multi-dimensional outcomes: socio-economic, cultural, political, humanitarian and physical. From natural disaster research, one can distil the following points:

- Natural events result basically from interactions between the atmosphere, hydrosphere and lithosphere; they are geophysical processes. Resource depletion and environmental decline change these interactions which, in turn, affect the frequency and magnitude of natural disasters (Tolba, 1992, pp.86-102).
- Natural hazards potentially threaten human society. Earthquakes would not be hazards if humans did not occupy active seismic zones. This means that natural hazards are generated from interaction between natural events system and human use system (Burton *et al.*, 1978).
- Natural disasters are the interaction between natural hazards and vulnerable conditions (socio-economic, cultural and political) which are usually created by human actions. Thus the distinction between natural and man-made disasters is blurred; many of the tragic impacts of natural disasters result from human misuse of resources, inappropriate actions and lack of foresight (Davis, 1987b, p.8).

- Natural disaster causes tragic impacts which disrupt established ways of life. Some impacts are quantifiable: loss of life, destruction to buildings and infrastructure, and economic losses. Others are less easily measured: loss of opportunities, changes in the community's culture, impact on the environment and socio-economic problems (UNDRO, 1991b, pp.106-112).

The interrelated, complicated nature of these points renders any manageable definition of natural disaster contentious and difficult. However UNCHS (United Nations Centre for Human Settlements) (UNCHS, 1994) offers a holistic framework for natural disasters, in terms of creation, effects, outcomes and responses (Figure 1.1).

> ... a natural disaster could be defined as the interaction between a natural hazard, generated in most cases from a sudden and unexpected natural event, and vulnerable conditions which cause severe losses to man and his environment (built and natural). These losses create suffering and chaos in the normal patterns of life, which lead to socio-economic, cultural and, sometimes, political disruption. Such a situation requires outside intervention at international and national levels in addition to individual and communal responses.

Such a definition represents a shift from focusing on the technical aspects of the physical phenomenon to a wider perception, that natural disasters are predominantly *environmental, social and developmental issues.* First, natural disasters are neither unforeseen nor isolated problems. They are integral parts of the spectrum of man-environment relations; this is an *ecological perspective*. Humans can indeed instigate natural hazards, directly and indirectly, by their inadequate or inappropriate intervention in the environment.

Second, the impact of natural disasters is affected by a society's values and institutions which shape the development and planning processes, the production of goods and distribution of resources, and the provision of opportunities and socio-economic facilities; this is a *social perspective*. Consequently these values and institutions influence the vulnerability of the people through land use, housing, infrastructure and emergency planning.

Third, the impact of natural disasters on human settlements results from the failure of the human use system, pre- and post-disaster, to protect people and their belongings. In the pre-disaster situation, this is directly related, among many other factors, to settlement growth, construction methods, and planning for mitigation and early warning systems. In the post-disaster situation, the impact is influenced by the effectiveness of relief and emergency operations, as well as the availability of resources to deal with the losses and chaos. In short, the impact of natural disaster is directly related to the level and type of development of a society; this is a *developmental perspective*.

## Global patterns of natural disaster

The frequency and magnitude of natural disasters differ considerably in types, locations and times. These variations are related both to physical characteristics and socio-economic conditions of the different regions. For example, population density, settlement planning and local conditions cause differences in vulnerability to natural disaster. However, in providing global perspectives of natural disaster one would face several problems such as deficiencies in the available data, lack of a common definition of natural disaster, and emphasis on physical quantifiable losses (deaths, buildings, etc.) rather than on wider, long-term effects (social, developmental, etc.). Berke (1995, p.3) emphases this point; Systematic record keeping on disaster losses is sketchy at best in developed countries, but it is almost non-existent in developing countries. Nevertheless, despite deficiencies in the existing knowledge, a few generalisations can be drawn.

Human and economic losses caused by natural disasters are enormous. During the 1970s and 1980s, natural disasters killed 3 million people, adversely affected 800 million more, and in 1990 alone caused economic losses of US $47 billion (UNDRO, 1991a, p.3). Broadly speaking, both physical damage and deaths have tended to rise in developing countries, whilst increasing economic losses in developed countries have been matched by a fall in fatalities. (Burton *et al.*, 1978; Department of Humanitarian Affairs, 1993, p.27). However, the impact of economic losses is disproportionally greater for developing countries, due to their scarce resources and low productivity. Possible reasons for the upward trend in loss levels for developing countries are the increases in social, economic and physical vulnerabilities.

Generally, natural events, as geophysical processes, are not increasing; but natural disasters are growing, due to the increased level of settlements' vulnerability, especially in developing countries. Therefore the frequency and magnitude of natural disasters have risen dramatically world-wide. The number of natural disasters and their economic losses have escalated from 16 events and US $10 billion in the 1960s, to 29 events and US $30 billion in the 1970s, to 68 events and US $93 billion in the 1980s. The average losses have risen from US $3.7 billion per year in the 1960s to US $11.4 billion in the (Tolba, 1992, p.86).

Globally the number of hydrological disasters (floods and cyclones), is greater than the number of geological ones, (earthquakes, volcanic eruptions and landslides). Between 1947 and 1981, there were 554 hydrological disasters compared to 208 geological ones. Within these two categories, floods were the most frequent (45 per cent), followed by tropical cyclones (27.7 per cent) and earthquakes (21.1 per cent). Although floods are the most frequent natural disasters, earthquakes and tropical cyclones cause the most fatalities.

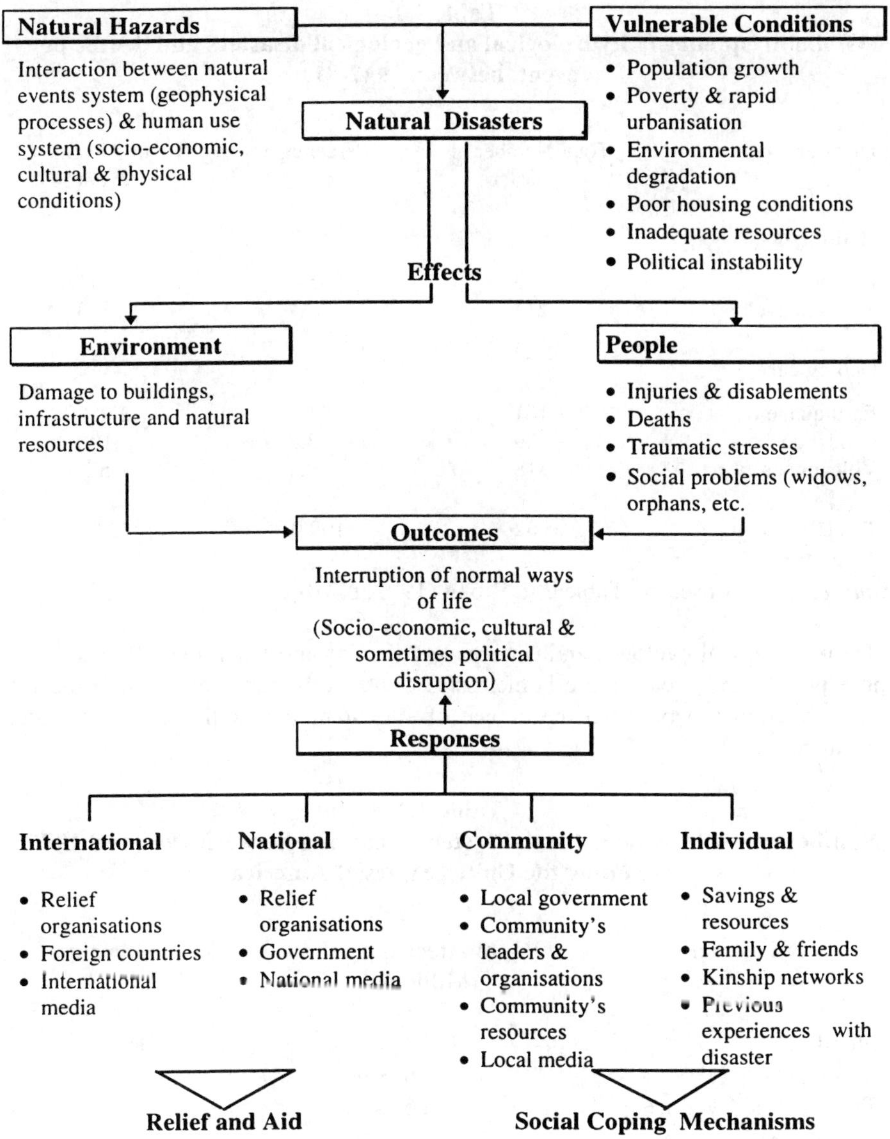


**Figure 1.1** **Natural disaster: creation, effects, outcomes and responses**

*Source:* Based on definitions in Burton *et al.* (1978); Davis (1987b); Krimgold (1974); Quarantelli (1978)

**Table 1.1**
**Global frequency of hydrological and geological disasters and deaths per event, between 1947-81**

| Disaster types | Total Number of Disaster | Percentage | No. of Deaths per event |
|---|---|---|---|
| Hydrological | | | |
| Flood | 343 | 45 | 571 |
| Tropical cyclone | 211 | 27.7 | 2,373 |
| Geological | | | |
| Earthquake | 161 | 21.1 | 2,652 |
| Landslide | 29 | 3.8 | 190 |
| Volcanic eruption | 18 | 2.4 | 525 |
| Totals | 762 | 100.0 | 6311 |

*Source:* Based on Table 2 in Blong (1992, p.210)

Despite tropical cyclones and earthquakes causing most fatalities, floods affect more people and create more homelessness. A likely explanation could be the strong relationship between the choice of site of human settlements and water resources.

**Table 1.2**
**Number of world-wide natural disasters occurring between 1900 and 1980, excluding the United States of America**

| Types of Disaster | No. of affected people (Millions) | No. of Homeless (Millions) |
|---|---|---|
| Flood | 339 | 36 |
| Earthquake | 26 | 10 |
| Typhoons & Cyclone | 26 | - |
| Hurricane | 3.5 | 1.2 |

*Source:* Based on Department of Humanitarian Affairs (1993, p.27)

Geographically Asia is the most disaster-prone area on the globe, as much of the continent is densely populated and lies in active seismic zones or near low-lying tropical coastal areas, subjcct to cyclones. Above all, the high degree of underdevelopment ensures that most of Asia has a large number of potentially vulnerable human settlements.

**Table 1.3**
**Proportional incidence of disasters and loss of life by continent, 1947-81**

| Continent | Disaster incidence (%) | Lives lost (%) |
|---|---|---|
| Asia | 38 | 85.7 |
| North America | 33 | 1.0 |
| Europe | 11 | 2.2 |
| South America | 6 | 4.2 |
| Caribbean and Central America | 7 | 4.5 |
| Africa | 3 | 2.0 |
| Australia | 2 | 0.4 |
| Total | 100 | 100.0 |

*Source:* Smith (1992, p.28) (adapted from Thompson, 1982)

## Urbanisation, poverty and natural disasters

In general natural disasters are increasing in number, scale and intensity world-wide. Nevertheless, their impact is clearly more severe in the case of developing countries, especially in Asia, as a result of precarious development processes including the lack of infrastructure to deal with disasters and the large number of people living on hazardous sites. Two important, interrelated factors increase the vulnerability of human settlements in developing countries: rapid and uncontrolled urbanisation, and worsening poverty.

*Urban growth*

Globally the growth of urban population has been astronomical. The number of people living in urban centres has risen from 730 million in 1950 to 2.3 billion in 1990, and is expected to reach 4.6 billion in 2020. A substantial 93 per cent of this will be in developing countries. Estimates are that 44 per cent of developing countries' populations will be living in urban agglomerations by the year 2000. In these countries, the number of urban inhabitants has escalated from 1 billion in 1980 to 1.4 billion in 1990, and is expected to reach 2 billion in the 2000 and 3.6 billion by 2020. These figures show that 2.2 billion people, between 1990 and 2020, will be added to the burgeoning cities in the developing countries - an increase of 160 per cent (Devas and Radoki, 1993, pp.1-2). Most urban population growth is not only the result of rural urban migration (39 per cent), more important is the natural growth of urban population (61 per cent) (Harphan *et al.*, 1988, pp.5-23). The resulting urban sprawl absorbs significant natural areas - in some cases valuable agricultural land. Between 1980 and the year 2000, urban areas in developing countries will more than double in size, from 8 million hectares to more than 17 million hectares (Tolba, 1992, p.188).

An important aspect of the population growth has been the phenomenal growth of the large cities, especially in the developing countries. In 1985, of the 10 largest cities in the world (those with more than 10 million population), 7 were in developing countries. Also in 1985, 25 cities in developing countries had a population of more than 4 million compared to 12 cities in 1970, and a further 35 cities had over 2 million. Between 1985 and 2000, these 25 cities are expected to accommodate an extra 100 million people (Devas and Radoki, 1993, pp.7-8).

This unprecedented urbanisation, rapid and uncontrolled, has been responsible for fragile, ad-hoc urban growth in developing countries. This is causing many environmental, social, health and economic problems. Choguill (1994, pp.23-28) discusses several underlying reasons for haphazard urban growth such as: (1) the weak institutional capacity in planning and jurisdictional conflicts; (2) scarcity and mismanagement of resources; (3) lack of cultural understanding; (4) inappropriate technology; (5) the exclusion of the target population from the planning and implementation processes; and (6) centralised top-down approach with its bureaucracy, biases and inequality.

Third World cities are increasingly becoming vulnerable to natural disasters as a result of the ways in which they have been built, expanded, serviced and managed. Most importantly urbanisation, as an interaction between human and natural systems, induces natural hazards and intensifies their impacts, through environmental degradation and disturbance of the ecosystem. This is one of the reasons why natural disasters cause more loss of life and damage to property in developing countries than in developed ones.

### *The scale of poverty*

In developing countries, urbanisation is associated with an increased level of poverty. An estimated 330 million city dwellers - 28 per cent of the developing world's urban population - fall below the poverty line. This percentage could increase to 57 per cent by the year 2000 (World Bank, 1989). Generally, rapid urbanisation and precarious economic conditions, with their national and international dimensions, have overwhelmed housing supply and infrastructure provision in most of the developing countries. Tolba (1992, p.190) estimates that 200 million people lived in slums and squatter settlements in 1970, rising to 450 million in 1990. Housing problems are not only the result of poverty and rapid urbanisation, but also the outcomes of the ways in which housing has been regulated, managed and produced within certain socio-economic circumstances.

Unrealistic building by-laws and high building standards, bureaucracy and inappropriate housing policies, and lack of resources and institutional frameworks are few of a complex causes which deny poor people access to adequate housing.

Growth of shanty towns and the proliferation of squatter settlements are common phenomena in almost all developing countries. 'Creeping cancers', 'slums of despair' or 'poverty belts' are some of the dramatic expressions to refer to the squalid conditions and misery of these settlements. However these spontaneous, unplanned settlements have become 'slums of hope' for millions of poor people to acquire 'shelter in an urbanised world'. The Global Report on Human Settlements (UNCHS, 1987) indicates the scale of the problems in developing countries:

- 40-50 per cent of the population live in slums and informal settlements;
- 30 per cent do not have access to safe water supplies;
- 40 per cent do not have access to proper sanitation;
- in most large cities, only a quarter of solid waste is collected by municipal authorities; and
- other services such as social, health, educational and transportation lag far behind ever-growing demand.

Such poor housing conditions have serious environmental, social, health and economic implications which, in turn, increase the vulnerability of human settlements in developing countries. Inappropriate construction methods and building materials, lack of planning and mitigation measures, and exploitation of resources and pollution are some of the factors. Tragically, the urban environmental decline aggravates infectious diseases, and diminishes the yield of nature, contributing to further poverty through its effects on people's health and lack of resources. All these limit their capacity to earn adequate income; they are trapped in a vicious circle which perpetuates their misery and poverty and which in turn serves to magnify the effects of natural disasters upon their lives..

**Urbanisation, poverty and vulnerability to natural disaster**

In developing countries, rapid urbanisation and rising poverty are creating an environmental crisis. Although cities are the centre of production and economic activity, many developing countries have failed to manage the impact of urban sprawl and poverty. Lack of housing, infrastructure and job opportunities, coupled with scarce resources and unrealistic policies, are threatening health, economic productivity, environmental quality, and social stability. Most importantly, urbanisation and poverty are increasing vulnerability of human settlements to natural disaster.

Urbanisation, as an interaction between human and natural systems, is extremely harmful and disturbing to the balance of the ecosystem. Rapid and large transformation of natural land to man-made artefacts is causing extreme environmental modifications: climatological, hydrological, geomorphological and vegetational. For example, urban growth affects the hydrological process, including changes in total runoff, flow characteristics and decline in water quality. As such it also affects erosion and sedimentation. Deforestation can lead to change in precipitation patterns which may result in flooding and land erosion. Global warming may change storm patterns and also may result in rising sea levels which affect coastal settlements where the major part of the world population is concentrated. Deforestation predispose land to flooding by removing vegetation which absorbs water. Urbanisation induces natural disaster by upsetting the ecosystem. The major reasons can be summarised:

- intense ecological modifications and various types of pollution;
- land degradation, desertification, deforestation and loss of biodiversity;
- encroachment and exploitation of natural resources;
- overemphasis on industrialisation, technological and commercial activities and ignorance of long-term effects.

Poverty is also exposing human settlement to natural disaster. Under the pressure of survival, poor people ignore essential environmental, technological and safety measurers. 'Housing for the poor' grows on marginal land: steep slopes, flood plains or near garbage dumps or industrial zones. These locations are favoured because of their nearness to employment, low commercial land value and small risk of eviction. Others seek shelter in the inner city in run-down dwellings lacking basic amenities. A graphic example of the interrelation between poverty and vulnerability is the 1985 earthquake in Mexico city; 10,000 people were killed, 50,000 injured and 250,000 left homeless. A high proportion of those were poor people living in densely-populated, multi-household rental housing. Caracas in Venezuela is another example; about half a million people live in squatter settlements on steep slopes which are subject to landslides, especially after heavy rain. Between 1980 and 1989, 266 landslides caused severe losses (Hardoy *et al.*, 1992, p.55 & p.91). Tragically, in such poor environmental conditions, the occurrence of a natural disaster can lead to other kinds of disaster such as the spread of epidemics resulting from contaminated water, poor infrastructure and lack of emergency services. In developing countries, poverty increases the vulnerability of human settlements as a result of complex factors which include:

- dangerous and marginal sites - steep slopes, flood plains, industrial zones;
- inappropriate construction methods and building materials;

- overcrowding and squalid housing;
- haphazard, uncontrolled settlement growth;
- lack of amenities and poor infrastructure;
- lack of disaster planning and mitigation measures; and
- lack of adequate systems for prediction, early warning and emergency.

This physical vulnerability of human settlements in developing countries is a manifestation of inappropriate socio-economic conditions and institutional incapacity. Poverty and urbanisation are two faces of the vulnerability coin. The complexity of these issues can be observed in Figure 1.2

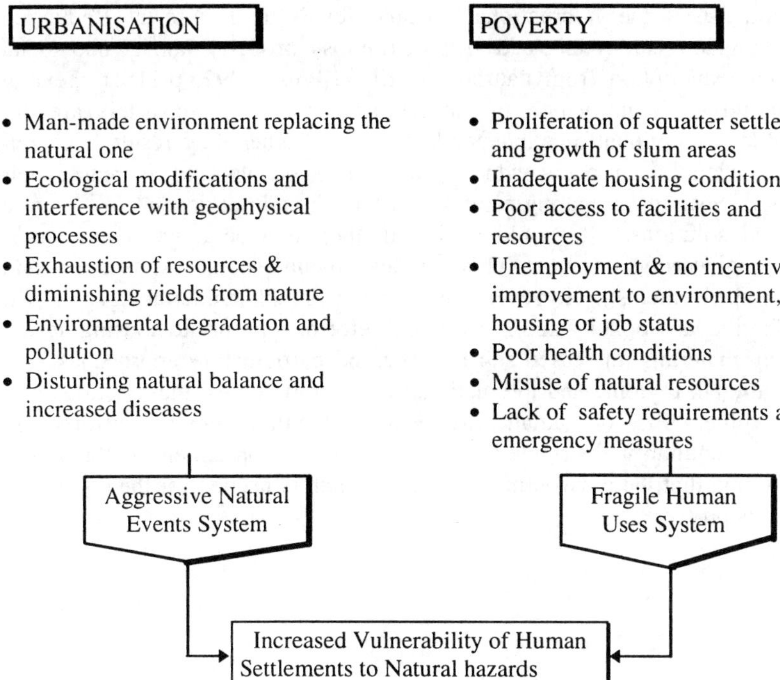


**Figure 1.2 Urbanisation, poverty and their effects on human settlements' vulnerability**

## Conclusion

No part of the world is risk-free from natural disasters, but developing countries are the most vulnerable because of rapid urbanisation and increasing poverty. These two factors are increasing the frequency and magnitude of natural disasters. Poor people are the victims of the socio-economic conditions which push them to live marginally on hazardous sites and in inadequate houses, as well as to seriously deplete surrounding natural resources. Such '*unsustainable*' growth of human settlements has created serious social, economic, health and environmental problems. It not only endangers the continuity into the future, but puts the existing built environment at extreme risk and wastes valuable scarce resources needed for development.

The importance of this matter has been recognised by the United Nations which declared the 1990s the International Decade for Natural Disaster Reduction (IDNDR). The aim is to reduce catastrophic life loss, property damage and social and economic disruption from natural hazards (Blong, 1992, p.218). Several lessons should guide mitigation of natural disasters. First, disasters are not unpredictable, exceptional and independent events; rather they result from on-going socio-economic process which marginalise people and increase their vulnerability. Second, mitigating natural disasters should not depend solely upon technological solutions. It should be based upon a wide range of measures including engineering solutions, land management, social adjustment and economic improvements. However this is not a simple undertaking. It will require effective land policies and institutional reform, appropriate housing design and building materials, increased self-reliance and participatory approaches, and adequate national, regional and international co-operation. All these actions will lessen the vulnerability of human settlements and will ensure that efforts and resources are equitably distributed and have lasting consequences for future generations, and that the environmental quality which is needed for the continuity of life itself is protected.

# 2 The planning process:

## An appreciation

*Michael A. Evans*

### Introduction

Planning is something we take for granted. Someone, somewhere is doing the planning. It is probably true that most organisations have plans of some sort and, when they work, they often go unrecognised. Yet there are also many instances where events have uncovered either the absence of plans, or their inadequacies. Planning costs time and money, so the question is begged - why plan? The simple answer is to look at what organisations or individuals have to lose by not planning. In disaster management, this is usually expressed in terms of responsibilities towards people, property and the environment. There is a simple test to help decide whether or not effort should be put into planning; it is to ask the question 'what if?'. If the answer is anything other than 'no problem', then some sort of plan is required unless of course, the attractions of a life littered with unpleasant surprises is overwhelming. Deciding that there should be a plan is the easy part, next comes deciding how to go about it.

There is no simple panacea, there is no such thing as a standard plan. Disaster management is a complex business involving predictability of risks, assessment of vulnerability, the extent of mitigation likely to be achieved and the balance of conflicting demands on time, effort and cash. No two situations are identical; for example, there are differences in the nature of the risks, in the inherent vulnerabilities, in resources and in cultural influences. There is however, a standard, analytical approach to planning, which follows a set structure and which will minimise the chances of missing some relevant aspect, or of failing to give due weight to a factor. That process is the subject of this chapter. It is straightforward and systematic to follow and it could make the difference between chaos and an organised, considered response.

### Who should do the planning?

In disaster management plans are needed at all levels, from government policy makers to grass roots. The depth, detail and often the time span of plans will tend to vary, depending on the planning level; for example, the higher the level of planning, the more likely it is to look longer term - often called strategic planning, - or to have more general application. The best planners are those who are likely to have to execute the plans. However, at more senior levels, it is likely that the planners will be at least one level removed from the 'coal face'. If so, it would always be ideal if they had first hand experience of the situations for which they plan.

### When to plan

The need for a plan can be created either by an anticipated or by an existing situation. In the former case, it would be described as a contingency plan - the classic 'what if?'. The main advantage of contingency planning is that it can be carried out without the pressure of time usually associated with urgent situations. It is unlikely that a contingency plan will exactly match the circumstances when they occur. This is not so much of a problem as it may appear, because it is much easier to adapt a good, well practised and disseminated plan to new circumstances than it is to prepare one from scratch and under pressure. This, in essence, is the difference between 'cold' and 'hot' planning. The former should always be the preferred option. However, the analytical approach described below can be applied to both cols and hot planning, all that varies is the depth and hence the speed of analysis.

### How to plan

One of the fundamental problems of planning is deciding where to start. A common method is to decide on a desired end state and work back. This may be perfectly valid for relatively simple problems, indeed it may often be the quickest way, but it relies on a high level of experience in the problem areas. Its main failing is that it presumes the end product; it is more of an intuitive approach. In more complex problems, the process must be more structured yet open ended, starting from a detailed analysis of the current or anticipated situation. It leaves as many options as possible open until its closing stages when the one most likely to succeed is selected. This avoids a tendency to massage the situation to fit a pre-conceived plan and, by following a format, the process helps to ensure that all factors and influences are taken into account.

The process moves in a number of steps, from a description of the situation, to defining the aim and then to considering factors which affect achievement of the aim. From there a number of courses which are open can be described, with each examined in terms of advantages and disadvantages. The course most likely to succeed is then selected. Thus the rationale behind a plan is argued and exposed. The process is called an 'appreciation'. Only when the best course has been selected can a plan be formulated. The plan itself should also follow a number of key headings, for example a clear statement of the aim, tasking, outline timings, logistics, communications systems and operational control arrangements.

A written appreciation has two main advantages:

- It clears the author's mind and it ensures that no relevant factors have been overlooked and that the course of action recommended is the best in the circumstances.
- It presents a clear picture of the nature of the problem to a higher authority, with the recommended course fully supported by valid and logical reasoning. This also provides a record of why decisions were made and forms a basis for updating.

**The format of an appreciation**

There are seven standard headings in a written appreciation:

- Review of the situation
- Factors affecting selection of the aim
- Aim
- Factors affecting achievement of the aim
- Courses open
- Selection of the best course
- Outline plan.

These headings should always be followed but the depth can be varied, depending on pressure of time. Throughout the writer should aim for accuracy, brevity and clarity. Detailed figures or tables should be included as annexes and should not clutter the main document.

**Review of the situation**

The review introduces the appreciation. It provides the essential background facts of the situation (either existing or anticipated) but it is not part of the argument. It serves three main purposes:

- It ensures that the reader is given an accurate picture of the situation.
- It shows that the writer has fully examined the situation.
- It provides a record of the basis of the appreciation including any assumptions which have been made.

It is often useful to make references to maps or charts which can be attached as appendices.

**Factors affecting selection of the aim**

A writer is sometimes given an aim, for example by a higher authority. If so, the writer must consult that authority in order to be clear on the intentions embodied in the given aim. The writer must understand what it is that he or she is expected to achieve. The subsequent analysis deals with how to achieve it. If no aim is given, it is in this section that factors affecting its selection should be examined. These factors may take the form of specific limitations, for example in terms of time or cost, or they could involve considering local market sensitivities, cultural influences, existing legislation or standards of training. Selection of the right aim is so fundamental to the appreciation that great care should be taken to get it right. However, it is not unknown for a selected aim to change during an appreciation. Occasionally during the later analysis, it may become clear that the aim selected is unworkable or unachievable. If so, a return to this section will allow for a re-examination and re-definition of the aim.

**The aim**

The aim is the focus of the appreciation. It is essential to get it right. It should be short, sharp, unambiguous and singular; it must not contain more than one aim. An aim is best written using a carefully selected infinitive verb. Weak verbs such as 'reduce', 'mitigate' or 'delay' should be avoided. It is also usual to include a short purpose in the aim, expressed in the form of 'in order to'. An example of an aim could be:

> To establish a contingency logistic capacity in the Great Lakes Region in order to be able to respond more promptly and effectively to unexpected refugee movement.

The aim may also be qualified by imposed, key limitations, for example by time. Once stated, the aim must be borne in mind throughout the remainder of the appreciation; all subsequent discussion must be related back to the aim. If there is no relationship, that part of the discussion should not be included.

There are four tests which can be applied to an aim, although they may not all be capable of application when the selection is first made:

- Does the wording say exactly what is to be achieved without saying how?
- Does it meet the author's instructions and responsibilities?
- Has it a reasonable chance of success?
- Is it the best available at this stage?

**Factors affecting achievement of the aim**

This section is where the main analysis takes place. Each disaster management situation will have its own set of factors but there are some which are usually present in all situations, these are: the threat(s), time, space, cost and resources. When deciding which factors should be considered, the test is to ask whether they affect the achievement of the aim in either a positive or negative sense. If they do not, omit them. When analysing a factor, the best way to do so is to make a series of statements of fact or reasonable assumptions and to draw deductions from them. The most powerful tool is to ask, after a given statement, 'so what?'. If the answer is 'so nothing', the statement was irrelevant and it should be excluded from the appreciation. The 'so what?' should also be applied to deductions until the answer 'so nothing' is reached. That will ensure that a particular line of reasoning has been followed to a full conclusion. It is usual to list factors starting with the most important. Among those which may be considered are:

- Range of possible hazards, their possible frequency and predicability or early warning (in essence, a hazard assessment). This factor can be taken as far as selecting the most probable hazard or the problem which is to be the main focus of the appreciation.
- The vulnerabilities of people, infrastructure, the environment and economy.
- Organisations and structures; government, non-government and community.
- Infrastructure; land, air and water route networks, alternatives, reliability, storage facilities, bulk handling.
- Resources assessment; internal, external assistance, transport resources; road, off-road, rail, air, river and sea.
- Other critical information requirements; finance, timings and distances.

- Communications systems, operations centre facilities or shortcomings.
- Legislation; enabling effects or constraints.
- Public awareness, social, cultural and political attitudes and constraints, community coping mechanisms.
- Other agencies involved; consultation, liaison.
- Climate, variations, effects.
- Emergency services; equipment and training standards, readiness, locations and facilities of emergency medical and health care.
- Sources of emergency supplies for each commodity likely to be required, order lead times and pipeline times.
- Utilities; water supply, sewerage, electrical power.
- Evacuation, mass casualties, refugee or displaced person facilities, temporary shelter, food, cooking and rest.
- Media.
- Rehabilitation, reconstruction.

Some of the factors inevitably influence others. Overlap is best dealt with where it occurs. To stick rigidly to compartmentalised factors would interrupt a logical flow and ignore obvious interaction. It is also important to make clear what is a deduction and what is not. The convention is to use the word 'deduction' as a heading. Avoid cluttering the arguments with detail which should be placed in appendices. When this section has been completed, and depending on the time available, it is often worth re-arranging factors into an order of importance which may not have been apparent at the start. Word processors make this a relatively simple task.

**Courses open**

In terms of disaster management, courses open usually relate to options for mitigation, preparedness, response or recovery. Whilst working through the factors, a number of courses open will have emerged. Avoid the temptation to invent courses simply to shoot them down; that would waste a readers' time. Concentrate only on those courses which could achieve the aim. Courses open can fall into 3 categories:

*Main Courses* Those which could achieve the aim - partially or fully.
*Combination Courses* A number of main courses which, when combined, could better achieve the aim.
*Complementary Courses* Options which are not capable of achieving the aim but which could enhance the achievement of other courses.

Each course should be examined on its own merits. No attempt should be made, at this stage, to weigh one course against another. Each option should be stated clearly and then discussed through its advantages and disadvantages. It is generally better to have a larger number of simple courses than a few complicated ones. The courses should be sufficiently different to warrant separate examination. An overview of a course can be added after each under the heading 'comments'. Courses are listed in the order main, combination and complementary.

**Selection of the best course**

It is here that the possible courses open are weighed against each other. This section forms the culmination of the arguments and analysis. It should be concise and direct. The comparison between courses should be developed logically so that the selection of the most appropriate is a natural conclusion. Some repetition will be inevitable but it should be kept to the minimum. No new material or ideas should be introduced at this stage. This section must finish with a clear recommendation of the course to be adopted.

The analytical part of the appreciation is now complete. However a little time spent checking the work will pay dividends. Check for logical reasoning, prejudice, relevance, ambiguity, accuracy and, most importantly, *will the selected course meet the aim?*. That done, the selected course must now be converted into an outline plan.

**Outline plan**

The outline plan serves two main purposes:

- It indicates the actions to be taken and so enables the reader to check that the selected course is workable, and is within the means available or predicted.
- It provides a mechanism for much quicker clearance with other agencies or higher authorities, so enabling progress to more detailed development without delay or nugatory work.

The outline plan must be a clear, practical proposal for action and the employment of available resources. It should be in sufficient detail (but only just) to convey how the selected course is to be put into effect. Others should be able to use it to draft detailed instructions, plans or orders. Once again, no new material should be introduced. If some were to come to light, the appreciation should be re-visited and the material discussed in the appropriate place. The outline plan should follow a logical sequence, usually chronological and it should include:

- A brief summary of the situation.
- The aim.
- A brief overview of how the plan will be put into effect.
- Task allocation of individuals, departments or agencies.
- Coordination arrangements including any key timings.
- Administrative arrangements
- Resource allocations including any external aid or assistance requirements.
- Control arrangements and communications systems to be used.

The appreciation must now (if time allows) be circulated for comment to all those involved. It is wise to set a deadline for comment or assumed acceptance in the absence of a reply. A consensus amongst those involved will clearly give the plan a head start.

**Review**

The useful life of an appreciation depends very much on the urgency against which it was written, but a contingency plan can have its usefulness extended considerably by review and re-appreciation. This can be built into the appreciation by adding a sentence at the end:

'Re-appreciate by..............' or 'Re-appreciate if.................'

The review tests to be applied are:

- Is the aim still valid?
- Is the reasoning still sound?
- Have any of the factors or their components changed?
- Have any new interactions come into play?
- Is the plan still workable and is it still the best way to achieve the aim?

**Was it worth the effort?**

If an appreciation sits in a filing cabinet gathering dust, the answer is clearly that it was not worth the effort. It must be developed into full plans which must be briefed and practised. The plans should be used as a basis for training the people involved. They could also be used for performance evaluation exercises or as a medium for testing new ideas. An appreciation is excellent practice in logical reasoning but is not an end in itself. Ultimately, it must stand up to the test of realities. Used properly, it will be one of the most powerful tools at the disposal of anyone confronted with a problem, whether they are disaster managers or not. Regularly practised, it can become second nature.

**Appendix** (Next page)

Framework of an Appreciation.

## Framework of an Appreciation

**An Appreciation of** ______________________________________

Identifying Reference
Date

For: By:
At: On: day, month, year

References: Maps, publications etc., listed and identified

**Review of the Situation**

1. 2. etc.

**Factors Affecting Selection of the Aim**

3. 4. etc.

**Aim**

5.

**Factors Affecting Achievement of the Aim**

6. Factor A
   f. Statements.
   g. Deductions.

7. 8. 9. etc.

**Courses Open**

10. Course 1.
    j. Advantages.
    k. Disadvantages.

11. Course 2. etc.

**Selection of the Best Course**

12. Comparison of courses open. It is therefore proposed to adopt Course x (or a combination of Courses y & z).

**Outline Plan**

Signature Block, Annexes, Distribution List

# 3 Reducing vulnerability following flood disasters:

## Issues and practices

*Dennis Parker, Nabiul Islam and Ngai Weng Chan*

### Disaster reduction and sustainable development in the 21st century

Two inter-related concepts achieved world-wide attention in the closing years of the 20th century, and look certain to be central to the disaster reduction agenda of the next century. The first is the International Decade for Natural Disaster Reduction (IDNDR). The Decade concept was adopted by the United Nations General Assembly in 1987 to draw attention to the global importance of more effective hazard and disaster management, and to highlight slowness and failure to incorporate many research findings into practice. In the same year the Bruntland (World Commission on Environment and Development, 1987) drew global attention to sustainable development as a criterion for wise choice about the environment. Disaster reduction and sustainable development are closely linked - this is especially evident in the Third World. The rising incidence of disasters throughout the world (International Federation of Red Cross and Red Cross Societies, 1993, p.33), is one of most commonly cited indicators of non-sustainable development (Mitchell, 1994). Many of the major disaster problems in the Third World are caused or deeply intensified by development problems, as in the case of drought and famine in the African Sahel. One of the principal ways of alleviating vulnerability to these disasters is to encourage economic and social development within sustainable limits.

Although societal processes are fundamental to human vulnerability to disasters the Decade has, in the eyes of some commentators, assumed the status of a 'technofix', in which those promoting technology and hard science use it as an argument for more of the same (Alexander, 1993, p.617). As Handmer (1995) points out, the division between those approaching hazards from a technical perspective and those approaching them from a socio-economic, institutional and political perspective, is part of the history of the field. Early and subsequent research on flood hazards by Gilbert White, first at the Department of Geography at University of Chicago (e.g. White, 1945) and later at the Institute of Behavioural Science at the University of Colorado in Boulder, was premised upon the need to approach hazard reduction from a multi-disciplinary perspective, rather than from a narrow engineering one, but this fundamental message has frequently been ignored. Through addressing the long-term effects of engineering structures on flood losses in the United States, and focusing upon the problem of progressive development of flood hazard zones, this research also took on an early sustainability perspective, questioning the sustainability of hazard-prone development and policies which encouraged it. The links between hazard reduction and sustainable development are now being addressed afresh through further questioning about, and distinguishing between, hazard reduction policies which are non-sustainable or compatible with sustainable development objectives (e.g. May *et al.*, 1996). This has led, for example, to Burton *et al.* (1993, pp.262-263) suggesting that sustainability, with respect to environmental hazards, should focus on reducing the risk of catastrophic losses, while enhancing the community's resilience to less dramatic events. Resilience is the opposite of vulnerability and needs to be cultivated and shaped to achieve both hazard reduction and sustainability.

This chapter does not set out to provide a comprehensive overview of the closely-to-be-linked fields of disaster reduction and sustainable development, although some further comments on their links are incorporated. The above introductory discussion is intended only to highlight the importance of certain issues including the 'technofix', the 'sustainability frame' and socio-economic, institutional and political 'contexts' of reducing vulnerability to flood disasters. Rather this chapter focuses principally upon (a) a framework for thinking about disaster vulnerability in the context of floods, and (b) practical problems and solutions leading in various ways towards resilience-building.

## Definitions, concepts and related issues

A combination of definitions and concepts is useful in developing a framework for thinking creatively about reducing vulnerability towards flood disasters in the post-disaster phase. However, it is instructive to begin by considering reconstruction and its relationship to other phases of disaster.

### *Reconstruction*

The reconstruction phase of disaster management is by definition a post-disaster one in which devastation and loss of life has usually already occurred. Although problems and issues exist which are more or less unique to the reconstruction phase, there are others which are common to several other phases (e.g. prevention, preparedness) and reconstruction. This is because disasters frequently occur in series and because it is in reconstruction that the seeds for future prevention and preparedness must be sewn. In the reconstruction phase there is often a strong desire amongst victims and their political representatives to return the damaged and scarred community to something approaching 'normality'. Here lies an initial problem because in many Third World communities normality is already an unacceptable condition for thousands, maybe millions, who suffer daily deprivation and risk which amounts to a continuous form of disaster (Kelly, 1995). In such circumstances reconstruction takes on a different meaning and relates to somehow propelling the damaged community upwards towards some new state characterised by a lesser degree of daily vulnerability and a greater degree of daily security and resilience: returning to normality is not an acceptable goal.

The most important aspect of reconstruction after disaster is that the beginning of this phase presents a host of new opportunities for governments, communities and individuals to avoid re-creating the conditions that led to disaster in the first place. Once settlements are established in hazard-prone areas, this 'window-of-opportunity' rarely presents itself, until a major disaster occurs permitting a fresh start. One exception is where a progressive urban renewal programme is seized upon to reduce hazard vulnerability either by removing development from the hazard-zone or by radically altering the design of new developments to enhance their hazard-resilience. Typically the onset of the reconstruction phase will present many new opportunities, only one of which will be hazard reduction. For example, developers are presented with opportunities to redevelop perhaps in a manner in which hazard-proneness may be increased (rather than reduced) since higher-density replacement development is likely to be profitable. In this context hazard reduction may take a back seat. The pressures to reconstruct following disaster may be intense as national and international commercial firms working in partnership with relief agencies seek to redevelop (Blaikie *et al.*, 1994, p.207). In such circumstances, there is sometimes little interest in seeking data (which often takes time to collect) to more adequately measure and define the hazard, and local knowledge is likely to be brushed aside as being 'unscientific'. Reconstruction to avoid re-creating the mistakes of the past is also limited by the informal economy, through which buildings may be reconstructed in situ without permit and to a low standard.

*Flood impacts*

The impacts of floods on society are complex, making them difficult to trace and measure. Floods are likely to affect both the formal and informal economies and, by definition, tracing the effects on the latter is problematic. Flood damages are usually categorised into tangible and intangible, direct and indirect and primary, secondary and tertiary ones (Table 3.1). Tangible damages are usually perceived to be those which can be measured in monetary terms, such as the damage to a dwelling, although such measurement is hardly ever precise and relies heavily upon estimation procedures. By contrast, intangible losses are those which either defy monetary measurement (e.g. the loss of an archaeological site through erosion caused by powerful flash flooding) or for which monetary estimates are considered undesirable and unacceptable. In some countries placing a monetary value on life is acceptable, while in others it is not. As monetary measurement techniques improve, losses previously regarded as intangible may become tangible. Direct flood losses are those caused by the physical contact of flood water with damageable property, whereas indirect losses are those caused by the consequences of physical contact of flood water with property. When floods drown livestock this is a direct loss, but when the income from livestock product sales is lost this is a consequential indirect loss.

Primary impacts are considered to be the 'first round' of impacts of flooding - the immediate effects. Typically these impacts lead to further 'knock-on' ones which are termed secondary ones. For example, the physical damage to a grain storage depot is a primary loss, whereas the food shortage which may follow is a secondary loss. A feature of models of flood impact, which attempt to include all categories of impact, is that they envision flood damages extending well beyond the area physically affected by the flood. This is because in practice, as with all disasters, floods set off a series of ripple socio-economic effects which extend well beyond secondary impacts and which economists may attempt to measure as 'multiplier effects' within the economy.

In reality the effects of flood hazards go well beyond the effects which flow from a severe flood event. Because hazard is an ever-present phenomenon in hazard-prone areas, where it and its potential consequences are well understood (perhaps because of the regular occurrence of extreme events), hazard may have a long-term negative impact upon the development process by presenting uncertainty and insecurity.

**Table 3.1**
**The typical range of flood impacts in a developing country**

PRIMARY IMPACTS

*Tangible, direct flood damage*
Damage to all building structures: homes, shops, manufacturing premises, public buildings etc.
Damage to the contents of buildings, including food stocks
Physical damage to infrastructure: roads, rail lines, bridges, water supply and sewerage plants and networks, electricity installations and power lines, hospitals and health care facilities, other communication networks

*Tangible, indirect flood loss*
Loss of industrial and agricultural production
Disruption of communications: by road, rail etc.
Disruption of utility supplies: electricity, water, sewerage
Disruption of health care services
Loss of crops and livestock

*Intangible losses*
Acute effects: mortality
Physical and mental impairment and morbidity
Destruction and loss of communities

SECONDARY IMPACTS (mainly intangible)

Food shortages
Increased mortality and morbidity
Homelessness
Shortage of clean water: reduced hygiene, contamination of water supplies
Disease epidemics: water-borne diseases, increased incidence of malaria
Disruption of health care and social services

TERTIARY IMPACTS (largely tangible)

'Knock-on' effects: refugees, migrations
Loss of exports; significant reduction of GDP

*Vulnerability*

In the human ecological approach to hazards (Burton *et al.*, 1993), interaction with the natural environment results in positive and negative outcomes. In the case of floods, the positive consequences of the interaction are income flows from resources such as flat and moist soil, while the negative consequences are threats to property and life and damages to them, as well as losses to environmental assets such as sites of special scientific or environmental importance and valued by humans. In the context of examining a household's vulnerability to floods, Green *et al* (1994), model household vulnerability as a function of (a) the socio-economic characteristics of the household (b) property infrastructure variables (e.g. the susceptibility of building fabric to damage) (c) flood characteristics (e.g. depth of flooding) (d) warning variables (e.g. warning time provided) and (d) response variables (e.g. number of people assisting). However, most definitions of vulnerability take a much broader view.

Hewitt (1983b) says that hazard refers to the potential for damage that only exists in the presence of a vulnerable human community, although this could be extended to components of the non-human environment valued by humans on the basis of the previous argument. However, the usefulness of Hewitt's conception lies in the idea that harmful events, which arise in hazardous environments (i.e. disasters), may be reduced by altering the vulnerability or the coping capacity of the exposed population.

The concept of vulnerability is the subject of intense debate and alternative definitions representing different disciplinary and ideological positions, and different end purposes. Blaikie *et al.* (1994, p.9) define vulnerability as 'the characteristics of a person or group in terms of their capacity to anticipate, cope with, resist, and recover from the impact of a natural hazard'. Environmental managers may criticise this definition for being anthropogenic, and those researching the vulnerability of infrastructure (e.g. food production systems and processes, roads and public utilities) to disasters have applied the concept of vulnerability to non-human, urban infrastructure to expose the human consequences of flood disaster (Parker *et al.*, 1987, p.33). For example, the vulnerability of urban economies to disruption from flooding has been modelled as a function of a number of key factors, as shown in the following formula:

$V = f(D, T, S)$ where

V = vulnerability to flood disruption

D = dependence - the degree to which an activity requires a particular good as an input to function normally (an example of indirect flood impacts)

T = transferability, the ability of an activity to respond to a disruptive threat by overcoming dependence, either by deferring or using substitutes or relocating

S = susceptibility, the probability and extent to which the physical presence of water will affect inputs or outputs of an activity (i.e. direct flood impacts)

In this formula, vulnerability is maximised where dependence is high, transferability is low and susceptibility is high, and is minimised by the opposite circumstances. For example, if food security is highly dependent upon bread availability and bread production is highly dependent on availability of flour *and* there is no substitute for, and no alternative source of, flour, *and* floodwater in the baker's premises is highly likely to stop ovens, then vulnerability is high. On the other hand, if food security can be guaranteed by alternatives to bread, or there are many sources of flour and baker's premises have been made resilient to floodwater penetration, vulnerability will be low. In reality, vulnerability of people and their support systems involves a large combination of physical, human, psychological and institutional factors which together determine the degree to which something or someone is put at risk (Blaikie *et al.*, 1994; Winchester, 1992b). Some of the principal factors contributing to high vulnerability are poverty, warfare and the denial of basic human rights.

*Types of vulnerability*

Several types of vulnerability may be distinguished which provide a broad framework for examining how vulnerability might be reduced in the post-disaster phase (Table 3.2). This framework demonstrates how vulnerability extends well beyond the initial and obvious effects of disaster. These types of vulnerability are similar to Foster's (1995) 'dimensions' of resilience: since greater resilience is what is achieved when vulnerability is reduced, the two concepts are highly complementary.

**Table 3.2**
**The principal types of vulnerability**

- Social vulnerability
- Institutional vulnerability
- System vulnerability
- Economic vulnerability
- Environmental vulnerability
- Vulnerability caused through unsustainable practices

High *social vulnerability* frequently promotes and exacerbates disaster and is often closely associated with under-development. Blaikie *et al.* (1994) provide a full analysis of the causes of social vulnerability which lie in social processes and differential access to resources. Although vulnerability and poverty are not synonymous, social vulnerability is often promoted by poverty and through related low response and recovery capacity. Social vulnerability may be reduced in a variety of ways. For example, Chan (1995) demonstrates how the vulnerability of poor, rural Malays to floods in eastern Peninsular Malaysia is much reduced by close kinship systems which exist in the flood-prone 'kampungs'. Rural Malays who migrate to flood zones in urban areas may have higher incomes, but often lose the support provided by kinship ties, which may make them more vulnerable than their rural counterparts. In seeking to analyse 'social resilience', Foster (1995) argues that a further form of social vulnerability exists. Under unstable political and social conditions in which power changes hands rapidly, social goals are frequently redefined and former achievements are re-evaluated as failures and vice-versa. In these circumstances Foster claims that resilience is best achieved by projects which have been designed to satisfy a diversity of social goals and objectives, so that when power shifts, current policies may be supported because they have relevance to the new regime. Resilience may also be promoted by accessibility to knowledge - widespread public understanding of policies and projects, but may be undermined by technologies which are perceived as complex, remote and threatening. Social vulnerability may actually be reinforced in situations in which there is little integration between formal and informal means of reducing hazards. For example, formal flood forecasting and warning systems imposed upon communities often have limited effectiveness, because the designers and operators of many such systems tend to ignore public preferences, and the potential offered by local knowledge and informal flood warning processes. Instead of attempting to find ways of integrating the best aspects of both formal and informal systems - to reduce social vulnerability - the common mindset is to supplant and replace the informal with the formal (Schware, 1982). Foster quotes a similar example of warnings of the November 1970 tropical cyclone in the Bay of Bengal being blocked in Bangladesh by officials, who were unhappy with the recent 'top-down' streamlining of the warning network, which contributed to at least two-hundred and twenty-five thousand deaths (Burton *et al.*, 1993; Foster, 1995, p.3).

*Institutional vulnerability* This is most the important cause of failure to effectively implement flood hazard reduction policies and strategies. For example, in analysing gaps in the implementation of hazard mitigation policies in the USA, Australia and New Zealand, May *et al.* (1996) identify institutional (a) capacity and (b) commitment, as two fundamental factors in determining the effectiveness of policy implementation. Where the capacity and commitment of implementing agencies is low, policy implementation tends to be weak. However, institutional resilience can be raised by enhancing capacity (e.g. the capacity to give technical assistance), or commitment to hazard-reduction objectives. Institution-building (e.g. introducing new legal requirements, capacity-building etc.) is therefore an important approach to reducing institutional vulnerability and increasing resilience to flood disasters.

*System vulnerability* This may be analysed in various ways using the vulnerability formula above. One approach is to reduce dependence on single sources of production or services. For example, instead of towns relying upon a single large water supply system, it is preferable to have a series of small-scale, self-contained water supply systems which offer the possibility of isolating damaged components and supplying flood affected areas from flood-free ones. Reliance upon simple, low-cost intermediate technologies to dispose of sewage may be preferable to more modern methods. For example, sewage removal by modernised cartage equipment for emptying household vaults, or bucket latrines, may be preferable to more sophisticated city-wide networks of sewerage systems, which may break down more easily in flood disasters' conditions (Parker and Thompson, 1991; UNDRO, 1982a).

Where possible, treatment plants, related installations and roads should be sited on relatively high ground. Here their exposure to flooding is generally lowest, thereby reducing their *susceptibility* to water damage. *Transferability* can be increased to raise resilience in many ways, usually by building-in a degree of system '*redundancy*'. For example, to reduce the possibility of food shortages in flood-affected areas, arising through their isolation owing to bridge losses and road closures, it is necessary to plan to store food in high locations within those areas likely to be affected, so that demands on food-carrying road transport to the affected area can be deferred in time. Similarly, constructing a road network with redundant linkages, some of which are over high ground, will reduce vulnerability to floods. Foster also identifies redundancy as one of the most important system characteristics (Table 3.2) required for resilience in the face of disaster. Following the Kobe earthquake in January 1995 the weakness of Japanese manufacturing systems was exposed: many manufacturers relied on a 'just-in-time' parts delivery system. As Foster (1995, p.6) points out, production is closely co-ordinated between plants, and components are provided as needed by a sophisticated, high speed transportation system. However, there is little redundancy in this system, which is *dependent* upon the continuous, efficient functioning of every component.

Resilient systems are also efficient and reversible ones. Inefficient systems commonly produce large amounts of waste, often producing a source of hazard too (e.g. hazardous waste), and are incompatible with environmental sustainability. Resilient systems also contain processes which may be readily halted and reversed. The organisational sociologist Perrow (1984), identified a number of problems which are endemic in the day-to-day operation of many socio-technical systems, such as advanced industrial production plants. These systems commonly display 'interactive complexity', which is unthought-of interdependence between components, and 'tight-coupling'. Tight-coupling describes the condition in which the failure of one system component leads very rapidly to the failure of others in a way which cannot be controlled, reversed or retrieved, such that an accident or disaster is bound to occur. Perrow categorised chemical plants, such as the one at Bhopal which caused disaster following gas leakage in December 1984, as complex and tightly-coupled.

*Economic vulnerability* This may reveal itself in various ways which promote disaster and/or a less than optimal response to hazard reduction. Lack of necessary analytical methodologies, knowledge and data associated with floods are all indicators that important hazard reduction investment decisions may be being made on the basis of poor information. An important advance in hazard reduction methodologies is to acquire sufficient data and modelling techniques to enable the capacity for benefits and costs of various mitigation options to be compared. This requires high quality data on hydrological and hydraulic parameters, damage potential (because damages which may be avoided are potential benefits) given events of different magnitude, and data on cost of mitigation projects. Economic resilience is likely to exist where disaster mitigation projects provide a high benefit-cost ratio and an early return on investment. Unfortunately, many large scale projects, such as flood control projects, do not display these characteristics according to Foster (1995). Often their costs escalate, leading either to abandonment and loss of investment, or they are completed with dubious viability. On the other hand a series of well co-ordinated, smaller projects provide greater flexibility (to change decisions) and less risk, thereby generating greater resilience.

*Vulnerability caused by non-sustainability* This has become a major cause of global concern (World Commission on Environment and Development, 1987). Resilient, sustainable development depends upon converging intra and inter generational equity, and upon the long-term sustainability of environmental systems. Environmental resilience exists where ways of life and disaster reduction projects are compatible with environmental sustainability. Many disasters, particularly those in the Third World, have a complex causation which importantly includes environmental degradation caused by misuse of natural resources. Overgrazing leading to loss of vegetation and soil erosion, and subsequently to desertification is one common example. Because they often affect the poor more than the rich, floods and flood mitigation projects may widen the income gap which is incompatible with goals of sustainable development (World Commission on Environment and Development, 1987). The challenge is to reduce floods in a manner which is compatible with reducing the income gap over time.

In the context of disasters, *technology* is a double-edged sword because it confers benefits to society and it also introduces risks or threat of harm. Sometimes technological responses to flood disasters, which are designed to increase resilience, actually increase vulnerability. Whether the net effect of a technological response to a flood problem is beneficial or harmful often depends upon the time-frame of the evaluation, and upon its geographical boundaries. For example, engineered flood control structures, such as flood embankments, may well reduce flood losses in a protected community in the years immediately following their completion. However, research indicates that such flood mitigation projects often induce further flood vulnerable development on flood plains, which in the long-term may lead to greater annual average flood losses (Parker, 1995; Thompson *et al.*, 1991). Similarly, the local effect of erosion and flooding prevention projects along the coast is often assessed to be beneficial, but when the spatial boundary of evaluation is enlarged, these projects are demonstrated to have significant adverse impacts in the 'downdrift' zone, often causing or contributing to coastal erosion and flooding problems there.

*Flood protection and residual hazard*

Floods can hardly ever be totally prevented yet flood protection projects often give the impression that the hazard has been removed. Where an area is protected by the construction of a flood embankment, the embankment will be built to a particular design-standard. This may, for example, be the 1:50 year standard which means that a flood of a magnitude which can be expected, on average, once in 50 years will be protected against, *providing* that the flood embankment is not breached (through inadequate construction or maintenance for example). However, the flood embankment will not totally protect against floods of a higher magnitude, or lower probability, than the 1:50 flood. Such floods present residual hazard which is often poorly understood and prepared for. Resilience can be increased by recognising and addressing residual hazard, perhaps through emergency preparedness measures designed to cope with the extreme breaching or overtopping flood event.

*Range of choice and vulnerability*

An important means of reducing vulnerability to flood disasters in the reconstruction period is to consider the full range of choice available for reducing future loss. The 'range of choice' concept was perhaps first introduced by White (1945), and its application is an important contribution to disaster decision-making processes. All too often key options for hazard reduction - particularly combinations of options - remain unconsidered and choice is rapidly narrowed, with the effect that disaster mitigation projects are less effective than they might be. This is particularly the case when single-disciplinary project teams are set to work on a disaster mitigation problem.

A critical problem afflicting the translation of the 'range of choice' concept into practice is limited definition of 'knowledge' of hazards. Taking a broader view can significantly increase the resilience of counter-disaster planning. Although attempts are now often made to utilise 'local knowledge' in disaster planning, there is at the same time a strong and common tendency to solely consider knowledge generated by the scientific community. Only that knowledge generated by 'experts' and 'professionals' interpreted as useful, and there is a great deal of difficulty in perceiving how local knowledge may be captured and used. Similarly knowledge from non-western, pre-industrialised societies tends to be over-ridden by knowledge derived from western industrialised society. However, as Handmer (1995) points out, there is now an increasing interest in 'traditional or local knowledge' and 'folk wisdom' (Correia *et al.*, 1994, pp.167-193; Parker *et al.*, 1994, p.164; Schware, 1982) and the use of consultative and participative methodologies to reveal it. Indeed, Wisner (1995) urges taking a wider view of the potential contribution which such knowledge can make to successful hazard reduction. He stresses that even when local knowledge is perceived to be useful, it is often only 'indigenous technical knowledge' which is perceived in this way. He argues that local 'social knowledge', for example about why hazards are perpetuated and in whose interest it is to perpetuate and reduce them, is valuable; as is local 'critical knowledge', for example, about why problems are either neglected or addressed at particular points in time.

## Vulnerability-reduction: examples and related issues

The above discussion suggests that there are numerous ways of reducing vulnerability and increasing resilience to future flood disasters. These should all be considered at the beginning of the post-disaster reconstruction phase, even if they have been considered previously. Indeed raising standards of infrastructure, housing, education, health care, insurance and poverty-reduction, through the general process of social and economic development, together with institution-building, should be considered as an important response to vulnerability reduction, even though these policies are usually more broadly targeted than at reducing hazard vulnerability per se. Each of the following examples was initiated in the wake of serious flood disasters.

### *Economic vulnerability and sustainability-building through an improved urban flood damage information network for Bangladesh*

Bangladesh is one of the most densely-populated and flood-prone regions of the world: for example, approximately 62 per cent of its land area was flooded in 1988. 'Normal' floods are an annual occurrence, but extreme flood events also occur somewhere in Bangladesh almost annually. Currently an estimated 24 per cent of Bangladesh's population is urbanised; a low percentage compared with other low-income countries such as India or Egypt. However, at seven per cent Bangladesh's annual rate of urbanisation is comparatively high, and the contribution of agriculture to Bangladesh's GDP is declining. Bangladesh's vulnerability to floods is currently mainly ascribed to the rural, agricultural areas which are receiving considerable flood mitigation attention under the current flood control strategy. In contrast, very few urban flood protection projects exist and much less is known about urban than rural, agricultural flood loss (Flood Plan Coordination Organisation, 1992). Vulnerability to floods is now rapidly increasing in the urbanising areas which are likely to present the greatest flood hazards for the future. A major challenge is faced by Bangladesh in addressing the growing flood vulnerability of urban areas, and finding a successful path towards environmentally sustainable development, given important questions over the compatibility of existing ways of life and flood control strategies with this goal (Westcoat *et al.*, 1992).

Given recent flood disasters in Bangladesh, progressively greater attention needs to be concentrated upon urban flood hazards and urban flood loss, in order to build knowledge and data networks which may be employed to enhance economic resilience. Research by Islam (1996) is concentrating upon this problem in order to construct reliable sets of potential flood damage data, which may be used to estimate the benefits of flood protection and, thereby, the economic effects of urban flood protection proposals. Only by generating reliable urban flood damage data can the necessary calculations be performed, to test the economic effects of alternative strategies and projects. In the absence of these data, efforts to allocate economic resources in order to achieve economic and social goals (e.g. economic efficiency, a redistribution of economic resources) are seriously hampered.

Research on flood damage potential is based upon detailed investigations of flood damages in 3 sub-areas of Bangladesh using 3 reference floods: Tangail (1988), Bahubal (Habiganj) (1993) and Khatunganj (Chittagong) (1991) In total, flood damage in 356 households, 123 industries, 103 other businesses and 20 offices, has been investigated by Islam. Table 3.3 presents total flood damage values for 4 principal urban house types in each of the 3 sub-areas. Per capita income is correlated with house type: per capita income is highest for house type BB and declines through BC and MC to the lowest which is MT. The potential value of these data lies in their incorporation within unit-loss flood damage models (Parker *et al.*, 1987), which are employed in estimating the potential effects of flood protection.

An important finding of the research into urban household flood damages (Table 3.3) is that low-cost houses (with their low-income inhabitants) are highly vulnerable to floods: the poorer the household is, the higher the percentage of damage to their asset values. These findings suggest that floods not only reinforce poverty but may also act to widen the income gap between rich and poor. Because flood damages may be counted as flood protection benefits, and because the rich usually generate more benefits than the poor, conventional economic efficiency cost-benefit methodologies tend to direct protection towards the rich and away from the poor. Unless checked, for example by more highly weighting the value of flood damages experienced by the poor, this problem can lead to practice which is counter to the principle of converging intra-generational equity, which is embodied within the concept of sustainable development.

**Table 3.3**
**Urban household flood damage estimates, Bangladesh,**
**(in 1992/93 Taka £ values)**

| House Type | TANGAIL RIVER FLOOD 1988 Total damage value | TANGAIL RIVER FLOOD 1988 Total damage as a % of asset value | BAHUBAL FLASH FLOOD 1993 Total damage value | BAHUBAL FLASH FLOOD 1993 Total damage as a % of asset value | KHATUNGANJ TIDAL FLOOD 1991 Total damage value | KHATUNGANJ TIDAL FLOOD 1991 Total damage as a % of asset value |
|---|---|---|---|---|---|---|
| BB | 23807 | 7.3 | 14069 | 6.2 | 65642 | 22.5 |
| BC | 24425 | 10.7 | 10781 | 11.6 | 92122 | 32.1 |
| MC | 16890 | 11.2 | 5500 | 30.7 | 76654 | 55.0 |
| MT | 11561 | 26.2 | 10714 | 43.2 | 34709 | 64.1 |
| ALL | 19198 | 10.2 | 10230 | 11.4 | 61908 | 33.6 |

BB Brick floor, brick wall
BC Brick floor, corrugated iron sheet wall
MC Mud floor, corrugated iron sheet wall
MT Mud floor, thatched wall

60 Taka = £1.00 Sterling

*Source*: Islam (1996)

*Reducing the physical vulnerability of dwellings and households through flood-proofing techniques in Hong Kong*

The rapid urban growth achieved by Hong Kong has generated a significant rise in the standard of living of its inhabitants, rapid population growth and the development of villages and towns in the New Territories, adjacent to the border with the People's Republic of China. There are approximately 100 villages within 10 kilometres of the border, many of which are located partly or entirely within river flood plains. Development of new towns in the New Territories has led to particularly rapid land use change, which is associated with the restriction of drainage, land filling and the reduction of flood plain storage and land reclamation which has reduced drainage gradients. All of these factors have exacerbated the flood hazard, and damaging flooding is an almost annual occurrence sometimes associated with typhoon conditions. Because living standards have increased, housing quality has improved and household durable goods are in higher ownership. Flood damages are therefore escalating and there is a growing intolerance of flooding. Various methods of reducing flood disaster potential are being employed in the Hong Kong New Territories, including river channel improvements, institution-building (e.g. new legislation to designate watercourses to bring them under planning control), improved maintenance and management of watercourses and improved flood forecasting and warning systems. The overall strategy has also included flood-proofing methods, both for villages and for selected households.

Flood-proofing is a set of techniques designed to increase the resilience of buildings, or clusters of buildings and related infrastructure, to flood damage (Laska, 1991). Systematic investigation of these techniques and their applicability originated from observed property owner self-protective behaviour - based on local knowledge and experience - within flood plains. In most cultures some occupants of frequently flooded flood hazard zones find ways of adapting their properties to reduce their vulnerability to flooding. Typical examples include the construction of steps or flood shields across doorways to prevent the ingress of water, or the elevation of houses. Both techniques can be observed in Hong Kong. The systematic evaluation of the range of feasible techniques (Table 3.4) subsequently led to their more formal categorisation and promotion, and to identification of the appropriate circumstances for flood proofing (e.g. Sheaffer, 1960). In Hong Kong bunding of small villages has already proved a valuable vulnerability reduction measure (bunding: protection by embankment).

Following serious flooding in 1987, 1988 and 1989 an impetus developed to investigate the applicability of flood proofing measures in villages and common house types in the Hong Kong New Territories (Parker, 1992). The feasibility of flood proofing is a function of (a) the type and condition of buildings and their typical contents, (b) catchment and flood conditions and (c) the socio-economic characteristics of householders. For example, dry flood proofing is not a recommended technique where river valleys are narrow and steep and flooding depth is likely to be in excess of one metre. In the case of the New Territories catchment conditions are conducive to this type of flood proofing because valleys are flat and wide and flooding is generally less than one metre in depth. In the New Territories a constraint upon 'contingent' flood proofing (i.e. non-permanent flood proofing which depends on the property owner implementing a pre-conceived contingency plan) is the rapidity of flooding. Therefore permanent methods of flood proofing are preferable, although contingent methods can work if linked successfully to flood warnings. In the New Territories villagers who are likely to be amenable to flood proofing are those who are owner-occupiers, those with some disposable income, those who have experienced multiple flooding and those for whom some flood proofing has already proved to be valuable. New Territories villages are of four principal house types; each type lending itself to particular forms of flood proofing (Table 3.5). For example, Type A is a 65 square metre house with concrete frames and concrete or brick infill, and concrete materials. Table 3.5 was prepared to provide a systematic guide to the feasibility and appropriateness of different methods of flood proofing, given the typical variation of house types, flood conditions and socio-economic circumstances in the Hong Kong New Territories. This guide was subsequently used to identify appropriate methods of flood proofing designed to reduce the vulnerability of selected flooding in villages in the New Territories (Table 3.6), and the methodology can be applied to most circumstances.

*Enhancing social and system resilience through improved flood warnings in Peninsular Malaysia*

Malaysia is a newly-industrialising country with a high economic growth rate and rising standards of living. These are amongst conditions in which floods and other hazards can be expected to be magnified, simply because rapid economic change can have dramatic effects on natural and social processes (e.g. upon the hydrological cycle and the rate of development in floodplain). Generally, the rather stable communities in the Peninsula's eastern agricultural region, and along the east coast, are well adapted to 'normal' floods; many of them are characterised by dwellings which are raised above the floodplain or above tidal flood zones on stilts of varying height, a further example of local knowledge being utilised to reduce vulnerability. In contrast, communities in the western part of the Peninsula, which is experiencing rapid urbanisation (e.g. in the Kelang River Basin in which Kuala Lumpur is located), are generally less well adjusted to floods. This is because of the hydrological effects of urbanisation (i.e. increase in flood frequencies and magnitude and in the 'flashy' nature of floods) which is creating new flood hazards, higher levels of population mobility (leading to loss of local flood experience and knowledge), and the steady loss of indigenous coping knowledge and mechanisms such as the stilt house. As floods become more 'flashy', flood warning lead time (i.e. the time between receipt of a flood warning and the onset of flooding) is reduced.

More effective flood warning systems can enhance social resilience to floods, by delivering the benefits of either advanced, or intermediate flood forecasting technology to communities. The range of choice of flood hazard reduction options can also be enhanced by introducing flood warning systems. However, research indicates that flood warning systems are only likely to be effective where the local knowledge and the needs of the community are well integrated into the flood forecasting and warning system (Parker and Neal, 1990; Schware, 1982). The same approach can also enhance system resilience by providing communities with a complementary mix of approaches to flood hazard reduction. For example, rather than relying solely upon flood embankments which may fail or be overtopped, the resilience of communities can be enhanced - and the risk of catastrophic disaster may be reduced - by adding a well-designed flood forecasting and warning system to cope with the residual hazard. Chan (1995) completed a survey of 618 flood plain occupants who had experienced recent flooding in four flood-prone areas of the Peninsula: Pulau Pinang, Kuala Lumpur (both western Peninsula locations) and Kelantan and Pekan (both eastern locations). This showed that 54 per cent of flooded respondents were not warned of flooding and the percentage of those flooded who were warned is lowest in the more urban western locations (Table 3.4). As many as 48 per cent of respondents relied more upon their own judgement than upon a formal flood warning system. Overall only one-third of flooded respondents received a flood warning, and 31 per cent of these receive their warning after the flood onset (i.e. with zero warning lead time). Normally the higher the lead time, the greater the potential for flood damage savings. The mean flood damage saved by flood warning was $705 Malaysian (the mean monthly income for Malaysian flood plain households is $917 Malaysian) (Chan, 1995, p.318). The potential for increasing flood damage savings - and thus enhancing resilience to floods - is high where flood warning dissemination systems can be improved and closely linked to local needs, in order to increase the proportion warned and the warning lead time. Research from

**Table 3.4**
**The principal methods of flood proofing**

| | |
|---|---|
| Property relocation | Removing an existing structure and resiting it on flood-free land |
| Property elevation | Existing buildings can be raised in situ, i.e. 'retrofitting'. New structures can be constructed above a minimum floor level, consistent with the flood potential and defined within planning controls |
| Bunding and walling | Individual buildings or small clusters of them may be flood proofed by constructing small bunds or walls around them: 'neighbourhood' flood proofing |
| Dry flood proofing | Structures that are generally impermeable to water may be modified to exclude floodwater by installing watertight barriers or closures to openings. Waterproofing sealants used on walls or floors can reduce permeability. Flood shields are an example. |
| Wet flood proofing | Within buildings flood damage potential can be reduced by alterations and by reorganising space use, so that the most vulnerable items are located in the least damageable place. Relocation of electrical circuits and outlets from floor level to waist or ceiling level is an example. |

**Table 3.5**
**Summary of conditions in which different types of flood proofing are likely to be most applicable/successful and inapplicable/unsuccessful**

| | Principal Methods of Flood Proofing | | | | |
|---|---|---|---|---|---|
| | Relocation | Elevation | Bunding | Dry Flood | Wet Flood |
| Village House Types | | | | | |
| Type A : Modern | 1 | 1 | 1 | 2 | 2 |
| Type B : Old | 1 | 1 | 1 | ½ | 2 |
| Type C : Small, old | 2 | 1 | 2 | ½ | 2 |
| Type D : Small, timber | 2 | 0/1 | 0/1 | 0 | 2 |
| Contents | 1 | 1 | 1 | 2 | 2 |
| Flood Characteristics* | | | | | |
| Shallow 20-100 mm | 0 | 1 | 0 | 2 | 2 |
| Moderate 100-1000 mm | 1 | 1 | 1 | 2 | 2 |
| Deep 1000 mm + | 2 | 2 | 2 | 0 | 1 |
| Low velocity less than 1 m/s | 1 | 2 | 2 | 2 | 2 |
| Moderate velocity 1-1.5 m/s | 1 | 1 | 1 | 1 | 2 |
| High velocity 1.5 m/s | 2 | 1 | 1 | 0 | 1 |
| Slow rate of rise | 0 | 1 | 1 | 2 | 2 |
| Moderate rate of rise | 0 | 1 | 1 | 2 | 2 |
| Rapid rate of rise | 1 | 1 | 1 | 1/2 | 2 |
| Short duration 1-3 hours | 0 | 0 | 1 | 2 | 2 |
| Moderate 3-24 hours | 0 | 0 | 1 | 2 | 2 |
| Long duration 24 hours + | 1 | 1 | 1 | 0 | 2 |
| Very frequent 1 + per year | 2 | 2 | 2 | 2 | 2 |
| Frequent 1 in 3 years | 2 | 2 | 2 | 2 | 2 |
| Less frequent 1 in 4 years | 1 | 1 | 1 | 0 | 1 |
| Unpolluted | 0 | 0 | 0 | 0 | 2 |
| Polluted | 0 | 0 | 1 | 2 | 1 |
| Little rainwater | 2 | 2 | 2 | 2 | 2 |
| Considerable rainwater | 2 | 2 | 0/1 | 2 | 2 |
| Socio-Economic Characteristics | | | | | |
| Owner | 2+ | 2+ | 2+ | 2 | 2 |
| Tenant | 2+ | 2+ | 2+ | 1 | 2 |
| Experienced with floods | 2+ | 2+ | 2+ | 2 | 2 |
| No flood experience | 2+ | 2+ | 2+ | 0/1 | 0/1 |
| Disposable income | 2+ | 2+ | 2+ | 2 | 2 |
| Low income | 2+ | 1 | 2+ | 0/1 | 2 |
| Non mobile | 2+ | 2+ | 2+ | 2 | 2 |
| Mobile | 2 | 2+ | 2+ | 0/1 | 0/1 |
| Previous flood proofers | 1 | 1 | 1 | 2 | 2 |
| Not flood proofers previously | 2 | 2 | 2 | 1 | 1 |

0 - Inapplicable/likely to be unsuccessful; 1 - Might be applicable and successful; 2 - Highly applicable and likely to be successful

* flooding characteristics (e.g. depth, velocity, etc.) are only specified in approximate, indicative terms

\+ where socio-economic characteristics are assessed as having no significant direct bearing on the flood proofing method they are specified as 2+

0/1 or 1/2 where applicability/successfulness is assessed as either 0/1 or 1/2 the condition is marginal or may be applicable/successful only in some cases

**Table 3.6**
**Suggested approaches to flood proofing selected new territories villages**

| VILLAGE | SUGGESTED APPROACHES | COMMENTS |
|---|---|---|
| Shui Tin Tsuen | Clearance, relocation and rehousing. | Approx. 20 households in metal squatter shacks. |
| Ko Po Tsuen | Mixture of clearance and relation, and dry flood proofing to be extended. | Extensive wet flood proofing evident. |
| Sha Po Tsuen | Dry and wet flood proofing, with possibilities of using neighbourhood fortress flood proofing for parts. | Squatter shacks recently cleared by landowner: one only remains to be cleared. Passageways unusually wide (1-6 metres). Seepage through house walls reported as a problem. Neighbourhood flood proofing using fortress concept likely to be problematic because of passageway widths. |
| Pok Wai | Neighbourhood flood proofing using the fortress concept or extend dry flood proofing or mix of both. | About 100 households reported as flooded in past year by village representative. |
| Chuk Yuen | Neighbourhood flood proofing using fortress concept. | Many houses built in 1970s and since - in sound condition. |
| Mai Po | Neighbourhood flood proofing using fortress concept. | Much dry flood proofing evident: concrete flood shields and steps to doors. Narrow passageways: 1-2 metres wide. |
| Tai Tau Leng | Extend dry flood proofing or neighbourhood flood proofing using fortress concept in parts. | Complicated drains making simplification and/or sump pumping essential. |

Note: The suggested approaches are the result of a rapid, preliminary village inspection only and in some cases extensive and detailed further investigation is required to confirm sources and routeways of flooding; to confirm the findings and to identify detailed and practicable flood proofing designs.

Europe indicates that there are some important ways in which flood forecasting and warning systems can be developed to make them effective (Parker and Fordham, 1996). Although there are some obvious and important physical and cultural differences between south-east Asia and Europe, many of these research findings appear to have relevance in Malaysia (Chan, 1995). Under-developed flood forecasting and warning systems are 'forecast dominated' i.e. investment and effort in the technology and procedures leading to a flood forecast is far greater than that which is directed towards disseminating the flood warning to those occupying flood hazard zones. In advanced systems this problem will have been addressed, and methods of timely and rapid flood warning dissemination will have been proven, and their performance regularly monitored. In rudimentary systems untargeted 'blanket' flood warnings will be characteristic but in advanced systems the dissemination of flood warnings will be much more targeted, aided by high quality flood hazard zone information. Effective advanced systems will be based upon public preferences for warnings and upon the integration of formal and informal warning systems, making the best use of local knowledge and willingness to be participate, including through the employment of local flood wardens.

*Resilience-building by widening the range of choice: floodplain management*

One of the most powerful forces propelling increased vulnerability to floods is economic development which channels people and properties into flood hazard zones. Exposure to flood risks is growing rapidly in many parts of the world, through this process sometimes known as 'floodplain encroachment'. Systematic data on floodplain encroachment is sometimes difficult to obtain, especially in the developing world, where development is rapid and land use mapping and related data-bases are non-existent or rudimentary. Research in the United States and England and Wales has, however, demonstrated the almost inexorable urban encroachment of floodplains (Montz and Gruntfest, 1986; Parker, 1995; White *et al.*, 1958). Elsewhere, for example in Bangladesh, rural floodplains have been 'invaded' by progressively larger numbers of inhabitants, each seeking a more secure livelihood.

The most common response strategy is a protective one. Flood embankments, river diversions and channel improvements are the classic forms of structural flood defence. In many parts of the world these remain the accepted and dominant approach, although serious doubts continue to be raised (Westcoat *et al.*, 1992). Beyond the USA, governments have been slow to recognise the crucial point, which was recognised by White (1945), that non-structural measures are also necessary. Progressive extension of structural flood defences, and progressive heightening of existing flood defences, tends to part of an 'escalator effect' (Parker, 1995) in which higher levels of security induce further flood plain development behind flood defences. Rising flood damage potential is commonly used to justify more extensive and higher flood defences and so on. The effect is to raise the potential for catastrophic flood loss in the event of structural defences failing (i.e. through overtopping, breaching or failure to close floodgates), and to progressively increase the maintenance and replacement costs associated with structural flood defences. The sustainability of this approach, in terms of mounting costs of maintenance and progressive replacement of high technology structures, especially in the developing world, is highly questionable.

The principal form of non-structural flood hazard reduction is regulation of floodplain development, although flood warning systems are a further example. The aim of regulation is usually to prevent undeveloped floodplains becoming unwisely developed and to constrain further development in already partly developed floodplains. Wise use of floodplains can be defined in various ways but it might include development decisions which minimise the risk of loss of life.

The concept of 'floodplain management' has reached its highest articulation in the USA (Association of State Floodplain Managers, 1990). Floodplain management embraces regulation of floodplain development through planning controls and through insurance-based incentives. However, it also embraces a wide range of flood hazard reduction options including flood proofing, flood warning, and flood information dissemination and education. Appropriate combinations of flood hazard reduction methods are selected for use in particular floodplains, employing the full range of choice of options and thereby enhancing resilience.

## Conclusions

Disaster vulnerability is growing over time owing to a host of factors, including (a) socio-economic and political forces which exclude and deny access to resources, and which reinforce poverty and the growth of the income-gap, (b) the perpetuation of environmentally degrading and non-sustainable resource use practices, (c) urbanisation and the spread of people and development into hazard zones and (d) the negative feedback effects of human activities on biophysical systems (e.g. greenhouse gases contributing to global warming and the consequent hazards of global climate change). In this context post-disaster reconstruction provides important opportunities to slow and to subsequently reverse the process of vulnerability growth.

This chapter provides a conceptual framework for thinking about how vulnerability to flood disaster - indeed any hazard or disaster - might be slowed and reversed, together with a number of examples. Because hazard vulnerability extends well beyond the immediate causes and effects of disasters, the scope for vulnerability reduction is enormous and extends into most aspects of the development process. An important theme is the need to identify paths of sustainable development in order to reduce the incidence of disaster and hazard-proneness. In turn this highlights the need to develop and utilise analytical methods (e.g. flood impact estimation methods), which are capable of identifying the distribution, as well as the economic efficiency, impacts of post-disaster reconstruction projects so that intra-generational equity issues which are fundamental to sustainable development are addressed. Another theme of importance is the role of local knowledge and experience in hazard reduction, and the need to build upon indigenous hazard-coping mechanisms which are all too often supplanted by external ones which prove to be ineffective.

**Acknowledgements**

The authors wish to acknowledge the ideas provided by John Handmer in developing this chapter. The authors also wish to acknowledge Binnie Maunsell Consultants and the Hong Kong Government, the British Council and Overseas Development Administration and the Universiti Sains Malaysia.

# 4 Development, vulnerability and disaster reduction:
## Bangladesh cyclone shelter projects and their implications

*James Lewis*

This chapter is dedicated to the memory of
Fred Cuny
who lost his life last year whilst on a mission to Chechnya.

### Introduction

The objective of this chapter is to examine the requirements for long term change, in order that conditions which have become the contexts for catastrophe can be modified. Such an objective is not easily identified, least of all achieved, when practical and popular attention is necessarily focused upon short-term measures for relief and for the immediate saving of lives which might otherwise be lost.

This chapter argues that development for vulnerability reduction does not have to wait for disaster to start it off. However invariably in practice and contexts of conflict are no exception, the necessity for immediacy of short-term action will take priority over the longer-term.

Nevertheless, this chapter focuses on the longer-term, with respect to those engaged in efforts for the short-term and in the hope that the recurrence of catastrophic pre-disposition might be reduced by a modified future.

### Development and disasters: a cyclical concept

Disaster, reconstruction and development, usually represented linearly, are in reality cyclical[1]. Though a disaster may commence a process that leads to development, development has invariably preceded disaster and had a bearing upon the extent and implications of the disaster that ensued - for better or for worse (Examples of vulnerability being made worse are numerous; e.g.: development for tourism and commercial agriculture (Jeffery, 1980; Lewis, 1981, 1984, 1987, 1994a, 1994b).

Some questions remain therefore, associated with the relationship between disasters and development (Cuny, 1983): What kind of development made things worse and, more constructively, what kind of development would have made things better - and will make things better in the future? More precisely; without waiting for disaster to start things off, what kind of development is required for disaster reduction?

Reconstruction and development, often expressed synonymously (especially in architectural and physical planning contexts), are not the same. Reconstruction is a component of development; although not an absolute definition, 'reconstruction' carries a heavy emphasis of the physical, whereas 'development' can more easily (and necessarily) include the social and economic (the same distinction applies between 'physical planning' and 'development planning').

Disaster reduction has to be widely multi-disciplinary, its' management a pervasive part of local experience, of context and survival; but disaster management has become separated from the development of everyday affairs that create vulnerability. The impact upon development of disasters is frequently deplored - but what of the impact of development upon disasters?

**What is vulnerability?**

Vulnerability is a pervasive socio-economic condition; it is the reason why the poor and disadvantaged are the predominant victims of disasters. The day-to-day vulnerable condition, comprising marginalisation, poverty and deprivation, is the context and cause of vulnerability to rarer and more extreme hazards; which themselves exacerbate vulnerability to the day-to-day condition in their aftermath - a cyclical process.

Natural phenomena have less to do with ensuing disasters than vulnerability previously accrued from the perceptions, institutions, policies and activities of society (Hewitt, 1983a) which in themselves are the context for development. It follows therefore that the actions of man in appropriate development identification and implementation, can intervene in this cycle, and conversely contribute to vulnerability reduction. Vulnerability assessments are being used as a starting point for disaster reduction; but if development itself was to become more sensitive to potential natural disasters and to equitability, vulnerability might be avoided in the first place.

Vulnerability is commonly interpreted as a physical state of exposure, related to location and quality of construction. Vulnerability becomes manifest when these locations in particular are seriously affected by storm, flooding or landslide, and when some poorer areas of construction sustain more damage than others. That there are external pressures which prescribe some of those locations and conditions, and that there are few options to their occupancy or quality of life, is acknowledged as being part of the relationship between vulnerability and development. To seek modifications to development so as to avoid such pressures, is one objective of vulnerability reduction in development.

Vulnerability, by which a community is physically at risk, is compounded by socio-economic and socio-political factors, affecting the community's capacity to absorb and to recover, by accessibility to resources and, at the domestic level, by the degree of 'defencelessness and the inability to cope with risk, shocks and stress' (Winchester, 1992a). Events which may contribute to that inability may accrue over time in the economy, loss of land or crops, ill-health, or death, all of which could be caused by local or large scale disasters, as well as contribute to their effects.

Thus vulnerability, as the pre-condition of physical exposure to the initial impact and its' immediate effects, is only a part of the overall pervasive and negative condition in which vulnerability to one thing of nessecity means vulnerability to another. Vulnerability to the effects of natural disaster, (not only to earthquake but to its socio-economic aftermath), is also vulnerability to conflict and to civil strife (not only to the bullets and 'machetes', but to the socio-economic consequences of deprivation, destitution and migration). The agent of disaster and the physical effects may be different, but the socio-economic effects are similar - homelessness, hunger and disease and increased vulnerability to subsequent hazards and dangers.

*Survival*

Survival in this vulnerable condition is possible but fragile (otherwise habitation would not exist); damage may make survival in the duration and aftermath of disaster or conflict, more difficult or impossible; but without survival and the restoration of some degree of comparative well-being, recovery is also either impossible or extremely difficult to achieve. The crucial development issue of survival has, in the past, been left to emergency assistance, applied in response to the event; whereas appropriate development can make more assured a self-reliant socio-economic survival by the accommodation and supply of basic resources necessary for life support. Survival in the first instance could therefore be more certain, and there could be a decreased demand for costly emergency assistance.

## Equitable preventive development

Amelioration of the perceived causes and reasons of vulnerability has to be equitably applied, not as a response to events as they occur. Equitable preventive development has the potential for reducing the socio-economic effects of natural disasters (and of some conflicts), and for diminishing the likelihood of their recurrence or continuation. Accrued conditions of invidious deprivation and perceived disadvantage may lead to animosity and conflict, often under cover of more obvious ethnic differences. Whereas equitable development can be deployed to assist the establishment and growth of pervasive and comparative well-being, there is more likelihood of ethnic differences being absorbed into an overall cohesive and gregarious social network.

Disaster definition depends upon an event which renders a population incapable of recovery without external assistance. If therefore, by a kind of development appropriate to reduction of compound vulnerability, it became possible to enable communities to better survive and to recover without external assistance - then disasters would be reduced.

There is always an epicentre and a periphery to disasters of all kinds; but it can never be known where or to what extent these will be - be it earthquake, cyclone, flood or conflict. If therefore, by the initiation of measures which may be considered to affect only a periphery, then by doing so, that periphery will be limited in extent and the disaster will have been reduced.

**Vulnerability management in development**

The reduction of vulnerability, as a crucial path to disaster reduction, has to be focused not only upon 'protection' by technology in building construction, nor in warnings and communications (UNCED, 1993), but in measures more to do with accessibility of social and material resources, and social participation linked to cultural expression and traditional knowledge and norms. Disaster reduction has to be made, inclusive of the enablement of human ecological adjustments in the activities of vulnerable people. To maintain resilience and self-reliance to counter the 'effects and implications' of disasters is vital, rather than only technological resistance to the 'events and forces' of environmental extremes.

Disastrous extremes are extensions of a prevailing, normal hazardousness. Small and local disasters cause conditions of vulnerability which contribute to subsequent larger ones. Ad hoc response only to extreme and rare disasters is fallacious; it is the recurrence of small scale frequent events that require most attention, in themselves as local experiences and as creators of vulnerability to the large ones. Further, the more frequent are events, the more normal they are - to the point where normality itself is the vulnerable condition. Thus, it is the more normal environmental condition, the enablement of survival in that condition, and the development of context and infrastructure at local levels that are required, not only global response to the more unusual, emotive or spectacular disasters (Lewis, 1987).

Complex emergencies in many countries have created contexts in which conventional interpretations of natural disaster relief and of peacetime development are inappropriate and unworkable. Desperate crisis-driven decision making, often in competition rather than co-ordination, needs to be replaced by systems for inter-relationship, integration and co-operation, with a view to the medium and long term as well as to immediate actions.

Distinctions between relief and development are best avoided, in the practical relationship of each modified in a linkage expressing circumstances, needs and capabilities, time and change. 'Continuum' suggests an over-simplified progressive linearity, not compatible with the disruptions of complex and continuing emergencies nor with the cyclical relationship of disasters with development; it also suggests that the onus for change is on relief, whereas both conventional relief and development may have to modify in unconventional circumstances for the over-riding objective of equitable vulnerability reduction. Reactive relief, integrated with proactive development, could assist the achievement of self-reliance in the long-term.

## Cyclone shelter projects in Bangladesh

There are or have been approximately thirty-five projects involving 24 funding (or self funding) agencies for the provision of cyclone shelters in the coastal 'high risk area' of Bangladesh; the result is the completion of one-thousand-six-hundred shelters (1996). A rough estimate of the total cost of these shelters to date would be in the region of £95 million (US$140 million), exclusive of the costs of professional and technical assistance. A subsequent project is currently anticipated, which will provide up to a further two-thousand-five-hundred shelters of various kinds.

### *Cyclones in Bangladesh*

In addition to their impressive and erratic frequency and normal combination of heavy rainfall and cyclonic wind forces, cyclones in Bangladesh are often accompanied by a storm-surge. Storm-surges are formed by the combination of low atmospheric pressure and the configuration of the Bay of Bengal which traps and concentrates the surge as it approaches land. The low lying plains of coastal Bangladesh offer least resistance to this catastrophic characteristic.

In the April 1991 cyclone, most of the destruction, and many of the one-hundred-forty-thousand deaths (Chowdury *et al.*, 1993), were caused by storm-surge (Haque and Blair, 1992) which is said to have reached a height of more than nine metres (Figure 4.1).

It is storm-surge that makes cyclones in Bangladesh a different experience to those in most other cyclone-prone areas. The hazard of cyclone force winds is secondary to that of storm-surge and it is this characteristic that governs the definition of High Risk Area (HRA) in coastal Bangladesh. 'High risk' is currently storm-surge flooding of one metre or more, the area of which is currently being refined. One difficulty in doing so is the dearth of topographic information; the other is to find reliable recorded storm-surge heights inland - the observation of storm-surge inevitably carries a high penalty!

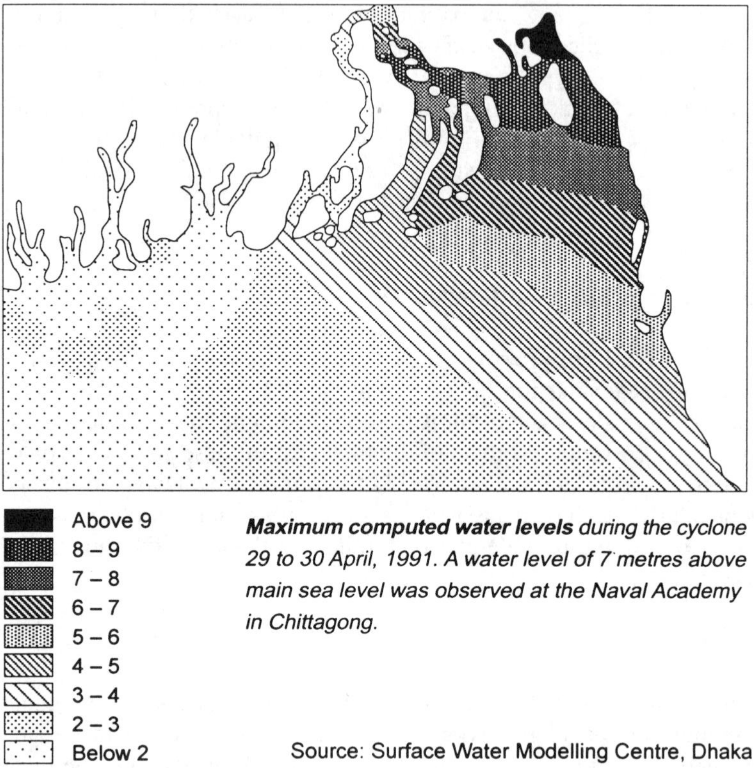


**Figure 4.1 Maximum computed water levels during the cyclone of 29-30 April, 1991**

*Multi-purpose and dual-purpose design*

Cyclone shelter construction commenced following the major cyclone of 1970 (approximately 358 shelters), and recommenced after the April 1991 cyclone. All of the pre-1991 shelters were 'multipurpose', designed primarily as shelters and taken over after completion for a variety of normal-time uses (i.e.: between cyclones: schools, colleges, etc.). Of the post-1991 shelters (BUET *et al.*, 1992), the majority so far (about 70 per cent) have been 'dual-purpose', designed specifically to serve a normal-time use. About three quarters of these serve as primary schools; others have been designed for unspecified normal-time 'community use'.

In the 1991 cyclone, many primary schools were destroyed or severely damaged in the areas of greatest cyclone impact (Districts of Chittagong and Cox's Bazar inclusive of the deltaic island of Sandwip and the offshore islands Kutubdia and Moheshkhali). Post-1991 provision of cyclone shelters has been concentrated in the Chittagong and Cox's Bazar areas. Where dual-purpose buildings (Lewis, 1994b) have been provided, this provision logically responds to the necessary replacement of primary schools. Where general community-use or multipurpose shelters have been constructed, this provision may not have been so logical, being a knee-jerk reaction to one particular cyclone, whereas the entire Bangladesh coastal zone is cyclone-prone.

More primary schools are required than just those needed for the replacement of those destroyed and damaged (the Government of Bangladesh has commenced provision for free education for all). Some dual-purpose shelters are being built on sites where there are existing schools (both then continuing to provide classrooms) and some are being built on new sites. The availability of sites for schools dominates site-selection for dual-purpose shelters. Where sites available to some programmes are restricted to certain uses (e.g.: primary schools), shelter provision in those programmes is denied to areas without sites (or without the need of schools).

*Building maintenance*

The principle advantage, from a donor's point of view, of a specific normal-time user are that the building is owned by an identifiable authority (e.g.: Ministry of Education or Primary & Mass Education Division) who could be expected to be responsible for building maintenance.

Arrangements for building maintenance are important because building destruction, in any natural disaster, occurs as a combination of building condition and natural forces. The pre-1991 shelters built in the 1970s, and most of the primary schools built in the 1960s, (destroyed in 1991), were in very poor condition. Concrete had spalled due to severely corroded reinforcement, brickwork was rotten and eroded, and door and window frames were missing or broken. Though the issue of 'ownership' may have been a cause of non-maintenance of the pre-1991 shelters, the schools destroyed in 1991 were the property of the Ministry of Education.

Reconstruction of primary schools to serve as cyclone shelters has necessarily been in reinforced concrete. Though not traditional, reinforced concrete construction in Bangladesh is nevertheless endemic. Maintenance however, though evident in traditional construction, appears not to be endemic. There are serious questions affecting project sustainability that remain to be settled, as to how maintenance (immediate and long-term) of non-traditional buildings is to be achieved. The issue has to be incorporated in the feasibility and design stages - not left until the building is completed.

*Construction cost*

Raised above ground-level to avoid and to resist storm-surges, built on clay/silt deposits (one third of the cost is in foundations), the cost of a primary school serving also as a cyclone shelter is five times greater than that of a normal primary school. Five new single-purpose primary schools can be built for the cost of one primary school/cyclone shelter. Is the difference the cost of saving lives?

## Reconstruction and vulnerability reduction

In 1991, embankments were in a poor condition and many were breached or washed away by the April cyclone. Embankment repair and reconstruction programmes are continuing; but as embankments succeed in reducing the storm-surge of cyclones, socio-economic vulnerability will be revealed that was once subsumed by physical failure and social catastrophe.

Post-disaster reconstruction is facilitating the development of primary education, as well as the anticipation of greatly reduced mortality in future cyclones. Similarly, dual-purpose cyclone shelter provision in the future is likely to be of benefit to other normal-time users in family welfare centres, mosques and 'madrassas' (teaching mosques), and market buildings etc.. Many vulnerable people now have a secure refuge to go to and many more will have refuge in the future.

Nevertheless, the physical success of cyclone shelter construction programmes, often against colossal logistic and environmental hazards and constraints, obscures some important social questions. Many vulnerable people may have a place to go, but what of reduction to their vulnerability?

Newly completed shelters are the object of vandalism in graffiti, illicit use of isolated roof spaces in otherwise crowded rural settlements, and difficulty with building maintenance; all these could suggest social rejection? Some Bangladeshi points of view consider the shelters to be socially divisive in their identification with a donor linked establishment. An elite who can afford to send children to school and contribute towards building maintenance are identified with that establishment, those who cannot are not. More invidious questions have emerged, concerning acceptability by the elite, of non-contributing poor and social outcasts (e.g.: single mothers) in the shelters in cyclones.

Some cyclone shelter projects have included the involvement of NGOs for the purpose of integrated community development programmes, especially for the formation and integration of women's groups and for disaster preparedness training. The time necessarily taken for social participation, community development and for the formation, for example of women's groups, by numerous NGOs of which some are long-established but some are not, is greatly in excess of that required to build a cyclone shelter. Contractually commencing at the same time, programmes for physical shelter provision and social community development quickly become out of step and left behind. Moreover, it cannot be assumed that all NGOs have the necessary resource capacity.

Social programmes for community development, riding on the back of technical assistance for shelter provision, are perhaps doomed at the outset. The alternative is for donors to run separate programmes or for different donors to work together on separate but complementary social and technical programmes, according to their expertise and experience. Is that too much to expect? It has been said that donors prefer the physical for the obviousness of the product; cyclone shelters can be seen and photographed, social programmes are not so apparent.

A more serious question is: what of socio-economic vulnerability? Will the physical technology necessary to resist massive natural forces, serve to displace the socio-economic development so crucially necessary for the reduction of socio-economic vulnerability? The physical success of the one will serve to expose the other. As more lives are saved, the quality of those other lives and their vulnerable condition will be exposed.

Vulnerability reduction depends upon the long-term establishment of social and economic back-up networks of and for community development; community groups, small credit availability, information and training. Additionally, what will be the impact of this massive infrastructural initiative upon education resources, community and social cohesion, NGO integration and programmes for women-in-development? What are parallel requirements, for example in health, agriculture, transportation, energy and communications?

**Architectural and social alienity**

Most cyclone shelters that have been built are urban concepts, totally alien to their rural contexts. As an expression of authority and power from afar, they do not express in any way the values of the rural user. As such they may be places that not only are not liked, but that are feared. They become used for illicit purposes because they are places which many people do not like to frequent (and such use increases their unpopularity).

Those who are able to use the buildings become an elite, related to the sources of power that the buildings represent (above), and are literally able to look down on those who are not. Small wonder that the building users express their satisfaction!

Rural rooms are usually arranged around an enclosed courtyard and always have a transitional space to pass through when entering and leaving; central circulation spaces are an urban concept. Wide verandas provide transitional space and create horizontality appropriate to a rural context. Building height should aim at being no more than tree height (two storeys). As much use as possible of natural materials would assist local affinity, and also serve to increase the possibility of local maintenance.

New shelter buildings now inevitably reflect the hazardousness of the Bangladeshi rural landscape, but in their alienage may be as much socially and institutionally out of harmony, as many are architecturally and environmentally.

## Strategic development for vulnerability reduction

The conventional linkage between disasters and development starts with a natural disaster and considers what needs doing so that subsequent disasters are less severe. There is nothing particularly wrong with this except that, whilst the process is continuing, other projects may be undertaken which are not part of those post-disaster initiatives. Whilst these specific initiatives are designed to make things better next time, other activities could inadvertently contribute to making things worse. There are cases where 'development' can be said to have been the cause of disaster; and there are countless examples of 'development' that has caused disasters to be more, rather than less severe.

An overall strategy across the development spectrum is required, which will exercise a co-ordinating role (and preferably a planning function for project identification as well), that will assess all development inputs, projects and programmes affecting disaster-prone areas on account of their impact upon equitability and vulnerability.

Strategic development for vulnerability reduction would consider the socio-economic, as well as the physical and infrastructural, dimensions. Physical and infrastructural development may be required to save lives (e.g.: embankments and cyclone shelters), but as more lives are saved the socio-economic dimension of those lives will be exposed and new questions will emerge: how will survivors stay alive, how will survival be transformed into recovery? What socio-economic/developmental inputs are required to facilitate self-reliant recovery? Save lives but provide for livelihoods as well.

In the essentially rural environments which are the context for this present chapter, development for vulnerability reduction would be multisectoral and consist of multiple small-scale micro projects for successful local integration. Physical and infrastructural development would include the provision of roads and pathways, essential for commercial and marketing activities (for an improved quality of life and well-being - the essential prerequisite for recovery), bridges, sluices and culverts (flood control and irrigation), rural electrification and energy generation, support for small industries and projects for agriculture, health and education. Selected and appropriate infrastructural input would have social benefits and moreover, in coastal Bangladesh, all 'pucca' buildings should be designed to double so as to be able to provide cyclone shelter facilities.

Social development would include community development, NGO integration, women-in-development, disaster preparedness training, preventative medicine and health services, maternity and child care, nutrition monitoring and small-credit schemes (Table 4.1).

**Table 4.1**
**Physical and social development for vulnerability reduction**

| PHYSICAL DEVELOPMENT (for) | SOCIAL DEVELOPMENT |
|---|---|
| roads and pathways, bridges | community development |
| sluices and culverts | NGO integration |
| rural electrification/energy generation | women-in-development |
| health and education | disaster preparedness training |
| markets | medicine/health services |
| food storage | maternity and child care |
| agriculture | nutrition monitoring |
| small industries | small credit accessibility |

DEVELOPMENT TO BE AVOIDED OR REQUIRING DETAILED SCRUTINY FOR IMPLICATIONS OF VULNERABILITY

Development for growth (only)
Large scale development projects requiring or causing migration and/or resettlement
Projects causing environmental damage and consequent loss of lifestyle and traditional resources
Inequitable development or allocation of resources which could cause or exacerbate cultural or ethnic divisions

By taking a spectrum of the objectives of development to be achieved in combination (not piecemeal alternatives) towards poverty reduction, the objectives of vulnerability reduction fit well and are significantly represented. Vulnerability reduction and poverty reduction are largely synonymous (Table 4.2).

The risk is that without an understanding of vulnerability and the need for its reduction in disaster prone areas, policies may ensue for limitation in the disbursement of national development funding - in the erroneous belief that less will be lost when disasters occur. The long-term effects of disaster impact and interaction with the potential for self-reliant recovery has to be understood for such beliefs to be eradicated and for the concept of equitableness to take its place.

**Table 4.2**
**Basic needs for poverty reduction**

| BASIC NEEDS | VULNERABILITY REDUCTION |
|---|---|
| Enough to eat | Yes |
| Clean water | Yes |
| A livelihood | Yes (logistics/marketing) small-credit) |
| A home | Yes (hazard resistant) |
| An education | Yes (information/training) |
| Health care | Yes |
| A safe environment | Yes |
| Protection from violence and natural hazards | Yes |
| Equality of opportunity | Yes (equitability) |
| A say in the future | Yes (development strategy/community development) |

*Source*: OXFAM (adapted)

**Note**

1 Created in 1943 for international relief and rehabilitation, the United Nations Relief and Rehabilitation Administration (UNRRA) was closely followed by the creation of the International Bank for Reconstruction and Development (World Bank) and the International Monetary Fund. This sequential division of responsibility is as current now as was arranged then - relief, rehabilitation, reconstruction and development. Recurrence was inconceivable at the end of World War II ('the war to end all wars') and the sequence was clearly conditioned by that one event. What was appropriate where recurrence was out of the question, was not appropriate where recurrence was in the natural order of things - and recurrence of wars did not end (Lewis, 1991) 'The relief syndrome: post World War II and sequentialism' Mimeo).

# 5 Learning from the people:

## A fieldwork approach in war-damaged villages in Lebanon

*Souheil El-Masri*

### Introduction

Understanding social context and assessing people's needs are prerequisites for planning reconstruction after disaster. What were the conditions prior to destruction and what are the people's resources? What are the factors which can enhance recovery and what are the opportunities for improvements? These fundamental questions must be answered to plan for comprehensive reconstruction. This is an approach to 'reconstruction with people' in which the social milieu of the disaster is understood and the victims need clearly articulated.

This chapter focuses on the methodological approach employed during the fieldwork in one war-damaged village in Lebanon (July-August 1990). The approach aims to gain insights into local conditions and the real needs of the victims. The chapter starts by highlighting the conventional mistakes made in assessing victims' needs in post-disaster situations, followed by a brief background of the case study. Then the weaknesses and strengths of each of the methods used in the fieldwork are discussed. It concludes with the main findings of the fieldwork which are based on a clear understanding of people's needs, aspirations, lifestyles and socio-economic realities. By highlighting these concerns, reconstruction and development could be integrated into a comprehensive framework.

### The myth of quantification

Experience with reconstruction after disasters has shown that the top-down approach, based on the intervenors' perceptions of what should be done and how it should be done, in isolation from the victims, has proved unsuccessful.

However the failures are not only limited to wasting resources, and not meeting people's needs, but extend to serious long-term consequences including the destruction of social coping mechanisms and increased dependency. These have obvious implications for the future development of the community. This approach, of 'reconstruction *for* people', which considers housing a quantified object, ignores the socio-cultural and economic aspects of dwelling (Figure 5.1). Unfortunately, this approach is still practised in many disaster situations. For example, the design entries of the 1990 competition 'Crisis Homes' organised by the International Union of Architects states:

> Hundreds of thousands of homes could be stored in boxes at airports in the West and flown out within hours of an earthquake or flood. The design needs to be simple so the home can be assembled quickly by unskilled labour, but robust enough to last years in extreme climates (Building Design, 1990).

Fieldwork in post-disaster situations, the subject of this chapter, has an essential role in planning for reconstruction, managing resources, organising activities and projecting alternative actions. However, experience has demonstrated that conventional assessment based mainly on quantification has commonly been ineffective. Cuny (1983, p.201) explains that:

> The problem is not that surveys are unnecessary, but rather the type of data that is most appropriate and the method should be used to obtain it... The loss of this time can mean loss of resources and commitment that would be invaluable.

The United Nations Disaster Relief Co-operation (UNDRO, 1982b) and Taylor (1981), among many others, have discussed the issue of a quantitative approach to assessing victims needs. They claim that the inadequacy occurs mainly in three areas:

- Lack of familiarity with the local conditions: housing conditions and processes, habits and lifestyles, people's needs and resources, household and location characteristics, etc..
- Inappropriate techniques: quantification of needs (counting the number of houses damaged and families affected); no distinction between needs and aspirations[1].
- Weak management: exaggeration, lack of a proper definition of the objectives of assessment, lack of co-ordination between different intervenors, assessment based on intervenors' self interest and available resources.

It is becoming clear that fieldwork in post-disaster situations should not only count things but it should define priorities, opportunities, problems and alternatives. Fieldwork should project the dynamic system of human settlement in all its aspects and should help in assessing the viability of alternative courses of action. Taylor (1981, p.139) argues that: Quantification is useful and may be necessary at a later stage when the needs have been hypothesised, but over-enthusiasm for the questionnaire as a tool often obscures rather than enlightens a situation.

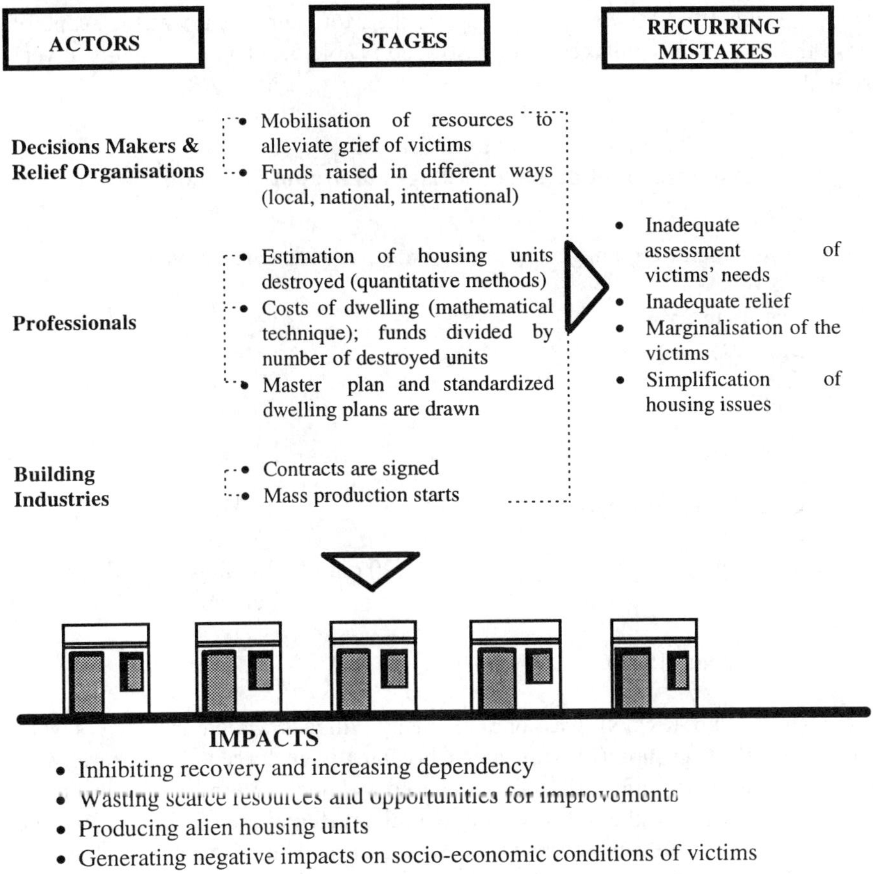


**Figure 5.1 Conventional approaches to reconstruction**

*Source:* Based on Cuny (1981, pp.3-9)

## The case study

During the sixteen years of hostility in Lebanon (1975-1991), warfare periodically raged, engulfing most regions of the country at one time or another. The haphazard terror left people confused with no logical explanation of the events which affected their livelihood. As Khalaf (1989, pp.20-21) explains: 'Recurring cycles of violence erupt, fade, and resurface again for no recognisable or coherent reasons'. From 1982-85, *as-Shuf* district, of the Mountain Area, was the scene of intense fighting. Many villages were destroyed (Table 5.1) and communities were uprooted.

**Table 5.1**
**Assessment of damaged villages in Mount Lebanon**

| Assessment of damage | Number of villages |
|---|---|
| Less than 25% | 10 |
| 25%-50% | 34 |
| 50%-75% | 16 |
| 75%-95% | 23 |
| Completely destroyed | 37 |
| Total | 120 out of 236 villages in Mount Lebanon |

*Source:* Faour (1988).

*Al-Burjain* (two towers) was one of many villages which suffered severe destruction. Both groups (Muslim and Christian) were forced, for a variety of reasons, to move to safer parts of Lebanon and the pattern of displacement developed according to the military actions in the different areas. After 1985, the Muslim group was able to return to *as-Shuf* area and three groups of families could be identified, during the fieldwork in 1990, based on their housing situation (Figures 5.2a & 5.2b).

- Families who were living in *al-Burjain* after finishing most of the reconstruction work on their original dwellings.
- Families who were living in *ad-Debbieh*[2], a village of evicted Christians which suffered damage, and who were in the process of rebuilding their houses in *al-Burjain*.
- Families who were living in *ad-Debbieh*[2], the place of refuge, and had not started rebuilding in *al-Burjain*, the original village.

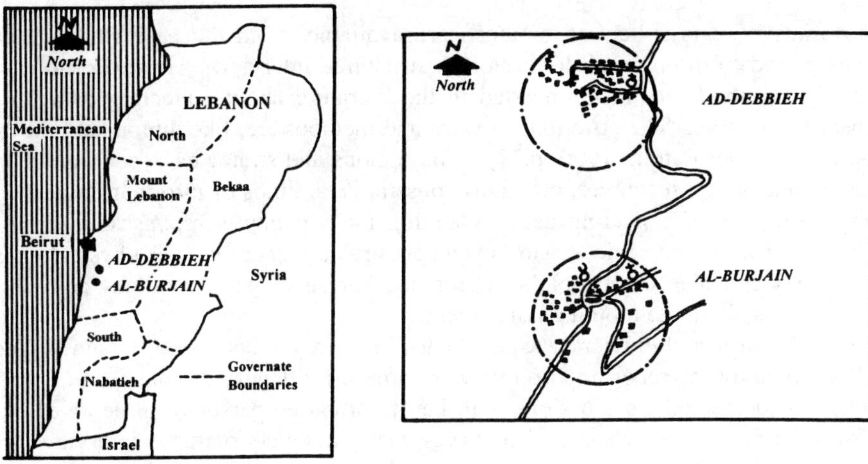


**Figure 5.2a Location of original village (*al-Burjain*) and place of refuge (*ad-Debbieh*)**

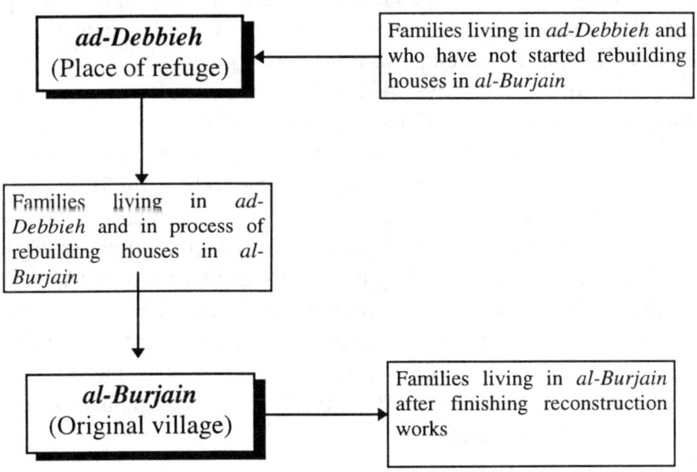


**Figure 5.2b Groups of families defined by housing situation**

**Fieldwork: methods and difficulties**

A variety of data-collection methods are available, from the case study to the census, and from informal discussion to structured interview. However, no rigid rules or methods are recommended in the literature about collecting data for a specific situation. The criterion for rules and methods are, like diplomacy, the art of the possible (Patton, 1990, p.13). The options and strategies depend on many factors including: resources, creativity, possibilities, kinds of information needed, and objectives of collecting data. Planning for meaningful reconstruction must consider a variety of socio-economic and cultural aspects as well as physical ones. It should consider the people's coping mechanisms, their resources and their perceptions about the rebuilt environment.

For this purpose, fieldwork was conducted in Lebanon, between July and August 1990, to study the reconstruction of war-damaged villages. It consisted of three interrelated methods: discussions with key figures (ten persons), in-depth family case histories (three cases) and a survey using a semi-structured interview (40 households). Despite the differences in these methods, they share important characteristics: flexibility in execution, deliberate interaction between the researcher and researched and a richness of data which stems from their largely textual nature and from their grounding in the language and experiences of the informants (Walker, 1985, p.7).

The sequence of these stages reflects different levels of focus on the issue of displacement and reconstruction (Figure 5.3). They are interrelated activities, one stage prepared the ground for the next. In fact, the development of a mixture of techniques capitalises on the strengths of each method and provides opportunities for cross-checking and for a complete picture of the situation being studied. This mixture of methods is imperative because there is insufficient information, especially official documents, about the village. The information has to be dug out as separate pieces from different sources (personal archives, newspapers, discussions, etc.) then assembled in a coherent form.

The political context and practical factors imposed many difficulties upon the fieldwork which was conducted solely by the author. In 1990, the fieldwork was postponed twice due to the unstable conditions in Lebanon. Also various aspects of the fieldwork were stressful and anxiety-laden. For example, 13 military checkpoints had to be crossed by the author travelling from West Beirut (place of residence) to the village; and 14 on the way back. The military situation and the sensitivity of some of the issues (causes of displacement, conditions for returning, financial aspects, etc.) required extra care and special tactics. Therefore, it was essential to develop communication methods for understanding the local context and assessing people's needs.

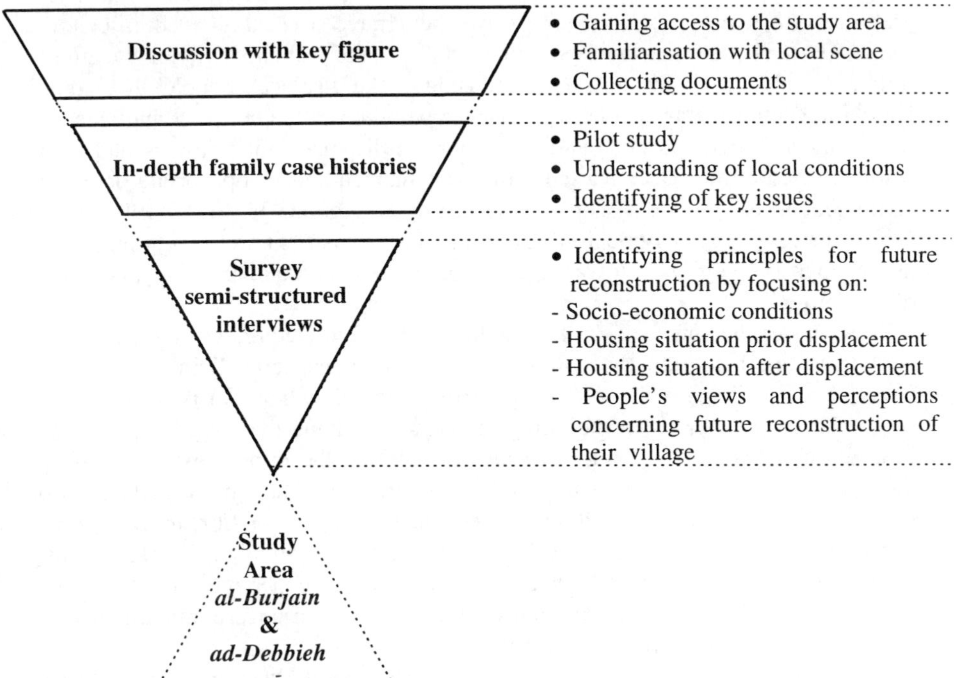


**Figure 5.3 Stages of fieldwork carried out in Lebanon (July - August 1990)**

### Discussion with key figures

Preparing the ground for the survey started with a series of meetings with the appropriate key figures. They are often referred to either as gatekeepers, those who control access to the study area and documents (Burgess, 1984, pp.45-48), or as key informants, those who have knowledge (stories, personal experiences, archives, memoirs, etc.), about the study area (Stacey, 1969, pp.47-48). However, there is no consensus on the borderline between a gatekeeper and a key informant; a key figure could be both (McNeill, 1985, p.68). The purpose of this stage is threefold: (1) gaining access to the study area, (2) familiarisation with the local scene, and (3) identification of key informants for discussions and references to be collected.

Gaining access to the study area is obviously a prerequisite for research to be conducted. The accounts of gaining access vary with the researcher, the researched and the research problem. Casley and Lury (1981, pp.130-131),

Burgess (1984, pp.31-52), Hershfield *et al.* (1983, pp.241-245), and Shaffir and Stebbins (1991, pp.25-81), among many others, have advised about this sensitive issue. This is particularly important in the case of Lebanon, given the highly militarised situation and the 'illegal' housing of the displaced people. The author had an advantage as an insider; a Lebanese who knows the culture, language and behaviour and who has personally experienced displacement. Access had to be resolved before the survey started; otherwise hurried and inappropriate decisions could jeopardise the field work. Concerning this point, McNeill (1985, p.68) says: The researcher will begin to penetrate the 'fronts' that are always put up for an outsider and will work towards developing a true and authentic understanding of the setting

The series of meetings snowballed, with one key figure acting as gatekeeper and leading to another, until the community itself was reached. Some key figures were from the village while others were from outside, with authority over the area or special knowledge about the village situation. Gaining the confidence of these key figures required 'negotiating tactics[3]' to create the opportunities to inform them of the intentions of the study and its importance. Thus in the early days of the project the author answered more questions than it was possible to ask, though in fact the key figures did prove more receptive than was anticipated. Despite this co-operation the author had to overcome high levels of personal anxiety at the outset and develop strategies to cope with the unease and uncertainty which were an inherent part of the project's execution.

The second purpose of this initial stage was familiarisation with the local conditions. Visits to the area and meetings with key figures helped considerably in getting a picture of the study area. However, familiarisation was a double-edged operation; the author had to get a picture of the area; the community had to get used to having a 'stranger[4]' in its midst. This process is:

> ... a social exploration. We learn what we can in advance about this relatively unknown territory, but once we are there, the first requirement is to gain some initial familiarity with the local scene and establish a social base from which we can continue our exploration until we are able to study some parts of that territory systematically. (Whyte, 1984, p.34)

During this stage, discussions with key figures were also used to identify and collect the necessary information about the village (discussions, personal archives, maps, photos, etc.) This was important as no complete data were available from a single source. To avoid wasting time and risk leaving without coherent information, a detailed checklist, which categorised and listed all information thought to be relevant, was prepared in advance. During discussion with key figures, relevant information was recorded in the appropriate category and the sources were listed. However, the checklist did not restrict the discovery of new information and data which emerged during discussion.

Discussion with key figures was carried out with both individuals and groups. Dealing with one key figure was easier and relatively straightforward as long as the topics for discussion were clearly defined. Group discussion was organisationally more difficult than individual interviews, which can be conducted in people's homes and at appropriate times. There are both advantages and disadvantages to group discussions. Group discussions are essentially social; they oblige participants to take account of other people's views in framing their response. Group sessions are dynamic and capable of stimulating discussions and generating new ideas, either by interaction with the members of the group or by listening and watching other peoples' interactions. There is more to react to, more food for thought, more diversity of opinion expressed than in a typical individual interview (McNeill, 1985, p.73). On the other hand, dominant figures can exert influence upon what is said; people may feel constrained in what they say in front of the others. This reaction was more evident when sensitive issues were discussed (political context, displacements, sectarian issues, etc.). Also group discussion offers fewer opportunities to explore individual positions in depth and breadth.

This stage, discussion with key figures, took time and effort equal to the main survey itself. But this investment proved to be beneficial in conducting the survey later on, and in anticipating some of the issues and problems at an early stage. It should be noted that a few meetings were held after the completion of the survey, either to report back for clarification or to collect documents which had been identified in previous discussions. Gaining access to the site and information was not a one-off event; it was instead an on-going social process that took place throughout the research project.

The unit of analysis consists of issues, ideas, and experiences and the reported and logical relationships between them. During the initial stage, an overview of the village (its socio-economics, building process, reconstruction issues, etc.) and verbal permission to conduct the fieldwork were gained. After spending some considerable time in the area, the author became aware of the community structure and the daily routine. This prepared the way for the second stage in which one family in each category was interviewed in depth.

### In-depth family case studies

The in-depth family case study method was used in the second stage of the inquiry; it in turn served as a pilot study to test the methodology and interview techniques to be used in the larger survey. It was also employed in an exploratory manner to gain insights into the situation and to identify an order of magnitude of key issues and thus aid prioritisation.

In analysing the experiences and attitudes of an individual, we always reach data and elementary facts which are not exclusively limited to this individual's personality; but can be treated as mere instances of more or less general classes of data or facts, and can thus be used for the determination of social becoming. (Plummer, 1983, p.64)

Three family case studies were selected to provide in-depth and detailed analysis about socio-economic profiles, general patterns of displacement, the building process, the problems facing the community before and after displacements and the people's perceptions of reconstruction. In this sense, the case study method serves as an 'equivalent to the spotlight or microscope' (Hakim, 1987, pp.61-75) involving detailed examination of a few families from which a great deal could be learned. This detailed examination is not only descriptive, but comments and alternatives are discussed when certain issues are highlighted to provide solid grounds for considering action in future reconstruction.

One of the principles of my work is to allow people to speak for themselves, to whatever extent this is possible, and in turn to communicate to them, in our conversations as well as in my writing, that it is their words I seek, and not material for the generation of something that ultimately transcends their words and hence their lives. (Cottle, 1978, p.xii cited in Plummer, 1983, p.84)

Finally, the stories and rich information of the case studies improve interpretation of the statistics (charts, tables, graphs, etc.) of the survey analysis. This provides better correlation between statistics and what it means in the life of the people in question, 'or, to use Malinowski's terms, it puts flesh and blood on the skeleton' (Lewis, 1976, p.3). This qualitative method is flexible in obtaining rich and in-depth information. However that same flexibility demands effective note-taking, recording, observing and conversational ability from the investigator. Yin (1989, p.26) supports this argument: Case study research is remarkably hard, even though case studies have traditionally been considered to be 'soft' research. Paradoxically, the 'softer' a research technique, the harder it is to do.

Despite the flexibility and the depth of the case study method, generalisation is often perceived as the major drawback in the rigous versus vigour debate. However, Yin (1989, p.21 & p.38) has argued that the aim of a case study is not to generalise the findings from a sample to a population by using frequencies enumeration (statistical generalisation), but to develop patterns and linkages of theoretical importance (analytic generalisation[5]). Generalisation could be achieved when the theory is tested through examining the findings in similar cases. Once such tests have been made, the results might be accepted for a larger number of similar situations (Yin, 1989, p.44). One should not claim generalisation, but special attention was paid to the selection of case studies to ensure their 'typicality' to the larger survey, which is the third stage of the field work.

**Survey by semi-structured interview**

Social surveys have been widely accepted by architects, planners and environmental designers as a basic tool for collecting data about people, their environment, and the interrelationships among them. Surveys are categorised either as qualitative or quantitative and each one has its strengths and weaknesses. There is no formula for choosing between the two categories; the choice is influenced by many factors: resources, accessibility, appropriateness, etc. Patton (1990, p.479) explains: The issue is not quantitative versus qualitative methods, qualitative methods are not weaker or softer than quantitative approaches - simply qualitative methods are different.

The choice of a qualitative type of survey was influenced not only by practical and economic factors. A qualitative approach is the best way to get in-depth details and rich information about a small number of cases. Using this method one gets rich details about the respondents' experiences and their socio-economic conditions, the ways they have organised their world, their thoughts about their built environment, and their perceptions about future reconstruction of the village. This is particularly useful for understanding issues in which processes and connections are important (visible and invisible structures of the built environment). It is imperative:

> ... to understand the process of housing and the invisible structures which shape those processes we need stories which correctly represent the world out there into which housing programmes intervene. We need, in other words, correct stories about process, about connections, and about the working rules of the housing system ... (Peattie, 1983, p.231)

This increases understanding of the cases and situations but reduces generalisability. Therefore the qualitative, non-probability, survey could not be grounded on established statistical techniques like the case of quantitative survey because of the way cases are selected. But validity and reliability could be enhanced during the sampling process to ensure 'representativeness' beyond the 'numbers game[6]'.

*Sampling*

Sampling is the procedure in which some part of a larger body is selected to represent the whole. Two issues are important in this procedure: the size and the way in which cases are selected. The sample size of the survey was 40 cases with a population of 334 persons and its basic unit was the household. Undeniably, the size was small comparing to the estimated population of the village - about 1716 persons forming 312 families.

> The validity, meaningfulness and insight generated from qualitative inquiry have more to do with the information-richness of the cases selected and the observational/analytical capability of the researcher than with sample size. (Patton, 1990, p.185)

Three essential factors have influenced the use of a non-probability sample. First is the lack of basic details (number of families, maps, etc.) from which a probability sample frame could be constructed. Second is the suitability of a non-probability sample to the type of inquiry which is based on a deep understanding of the processes, issues and perceptions. Third is the practical considerations to be taken account of in conducting the interviews in the study area. However, the selection of the sample units was not haphazard, as a practical sample framework was developed to identify the cases. The development of the frame was guided by the information gained during the first two stages of the fieldwork: discussion with key figures and the in-depth family case studies. The criteria for selection were the categories of households who represent different housing situations (stratified purposeful sampling). In each category, further sub-criteria were set up to ensure a complete and comprehensive picture of the situation. The sample framework was established to enhance validity and credibility by covering different situations and conditions This framework represents the grounds for the selection of the sample units to cover a range of:

- locations of dwellings such as: main road, secondary road and footpath;
- physical conditions of the dwellings such as: good, acceptable and bad;
- family groups; therefore surname used as a means for selecting the cases; and
- respondents from the households interviewed (father, mother, children).

The survey is the type of qualitative interview which aims to understand people's terms and judgements, and to discover the complexities of their individual perceptions and experiences. The main assumption behind the qualitative interview is that the perspective of others is meaningful, knowable, and can be explicit. We enter the field without predetermined phrases or categories which might constrain respondents' answers. Hence, such interviewing aims to capture what is in someone else's mind and to find out things which we cannot directly observe.

> The fact of the matter is that we cannot observe everything. We cannot observe feelings, thoughts, and intentions. We cannot observe behaviors that took place at some previous point in time. We cannot observe situations that preclude the presence of an observer. We cannot observe how people have organised their world; the words and meanings they attach to what goes in the world. (Patton, 1990, p.278)

This requires skill and experience from the interviewer and most of all the ability to listen when 'knowledgeable people[7]' are talking. Patton (1990, pp.280-290) lists three variations of interviews to collect qualitative data. Each has strengths and weaknesses and differ in the extent to which interview questions are determined and standardised before the interview occurs. The approaches are:(1) the informal conversational interview, (2) the general interview guide approach (usually referred to as semi-structured interview) (3) the standardised open-ended interview.

The second approach was chosen because it suits the type of investigation which combines the positive aspects of the other two. A semi-structured interview consists of a list of issues and questions that are to be discussed during an interview.

> An interview guide is prepared to make sure that basically the same information is obtained from a number of people by covering the same material (standardised character ). The interview guide provides topics or subject areas within which the interviewer is free to explore, probe, and ask questions that will elucidate and illuminate that particular subject. Thus the interviewer remains free to build a conversation within a particular subject area, to word questions spontaneously, and to establish a conversational style (conversational character) - but with the focus on a particular subject that has been predetermined. (Patton, 1990, p.283)

The technique is somewhat systematic because it covers the same issues for each respondent; this facilitates, to a certain extent, recording and analysis. However, it is not important that questions are asked in the same way or in the same order for all respondents. The interview can be developed in more or less detail, depending on how much the researcher is able to specify important issues in the course of the interview.

The structure of the semi-structured interview was adjusted and refined following the first two stages of fieldwork. The main issues explored during the interview were:

- socio-economic conditions of the people;
- housing conditions before displacements: process of building materials used; deals and arrangements, design aspects and considerations, problems and constraints, etc.;

- displacements and housing provision; and
- perceptions about house/village reconstruction: relation between physical and non-physical reconstruction, actions needed, alternatives, possibilities, etc.

The semi-structured interview was completed with relevant data obtained during the interview and supplemented by notes, observations, photos and sketches. Observations add important information about issues discussed. A skilled interviewer can read non-verbal messages and note reactions of the interviewee. Sharp observations can identify photographs to be taken or sketches to be made and which help in explaining the points under discussion (Yin, 1989, pp.91-92).

*Conducting the survey*

In the first period of the survey, the president of the village committee, who accompanied the author, recommended the selection of the cases. It was noticed that the president tended to identify cases of similar surname (his family). However, after a few days the president, due to time constraints, issued a written permission allowing the author to conduct the survey. Using 'the social base', established in the first two stages of the fieldwork, the author was able to identify a key figure - the vice-president's 16-year-old son - to accompany him. A local person was crucial for introducing the author to the households and for facilitating the work. This was a turning point in which complete control of the survey - flexibility in terms of time, and freedom in the selection of cases and discussions - began to be mastered by the author.

## Findings and conclusion

Analysis of the fieldwork has highlighted many important issues which should offer guidance to policy makers and professionals in planning for post-war reconstruction in Lebanon. Moreover, these issues contradict many of the conventional assumptions about victims' abilities to help themselves, building processes, rural life and reconstruction programmes. Some of the broad results can be summarised as follows:

- People have considerable ability to deal with the problem of their own homelessness, in contradiction of the popularly held perception of victims' helplessness.
- There is a high degree of changes in the socio-economic conditions, as opposed to traditional rural ways of life.
- People control the building processrather than public, centralised housing authorities.

- A wider concept of reconstruction (developmental, economic and social) emerged, as opposed to the solely physical reconstruction offered by top-down approaches.

Despite limitations and difficulties, the methods used in the fieldwork proved to be efficient, flexible and manageable. It provided a rich insight into intentions, feelings, relationships, contradictions, views and possibilities. Understanding local conditions and people's needs and aspirations was very beneficial in tackling the complex aspects of reconstruction (physical and non-physical) and in overcoming the obstacles of access to the site and to information. The fieldwork approach, discussed here, provides a clear understanding of the conditions of the community prior to and after the destruction of the village. It also develops an understanding of the conditions of the people, and their needs and perceptions about rebuilding their village. Such an approach promotes reconstruction which capitalises upon people's initiatives, maximises the use of available resources and enhances socio-cultural identity. In this way, reconstruction becomes a vehicle for development, which resolves basic problems, improves conditions, speeds recovery and encourages self-reliance.

## Notes

1 Needs and aspirations: Despite the recognition that assessment of needs must be determined from the victims' perspective rather than from donors' interests, an interviewer should also be able to draw distinction between aspirations (what people dream for) and the real needs (the normal and the appropriate and necessary to sustain life).

2 *Ad-Debbieh*: During the upheaval in *as-Shuf* area, many villages were damaged on different scales. *Ad-Debbieh* was one of the lucky cases which escaped complete destruction. Therefore safe houses there were the refuge for people from *al-Burjain*.

3 Negotiating tactics: This expression was used by Shaffir and Stebbins (1991, p.73). The uniqueness of each setting requires different tactics (friendliness, humour, sharing, networks, contact). This depends on the researcher's ability to engage in sociable behaviour which secures the access to the setting.

4 Stranger: In general, the people of a Lebanese village are close to each other. This closeness is based on strong blood ties and kinship structure. Hence, it is very easy for an outsider to be identified, who is then referred to as a stranger. For example, a man from a neighbouring village married a woman from *al-Burjain* which became the place of residence of the family. After 20 years, the man is still referred to as a stranger.

5 Analytic generalisation: Yin (1989, p.44) explains:

> This approach is well illustrated by Jane Jacobs in her famous book, The Death and Life of Great American Cities (1961). The book is based mostly on experiences from New York City. However, the chapter topics, rather than reflecting the single experiences of New York, cover broader theoretical issues in urban planning, such as the role of sidewalks, the role of neighbourhood parks, the need for primary mixed uses, the need for small blocks, and the process of slumming and unslumming. In the aggregate, these issues in fact represent the building of a theory of urban planning.

6 Numbers game: Patton (1990, pp.478-479) writes about this point:

> It is ... helpful to understand the special seductiveness of numbers in modern society. Numbers convey a sense of precision and accuracy even if the measurements that yielded the numbers are relatively unreliable, invalid, and meaningless ... Thus, by knowing the strengths and weaknesses of both quantitative and qualitative data, the evaluator can help stakeholders focus on really important questions rather than, as something happens, focusing primarily on how to generate numbers.

7 Listening to knowledgeable people: Yin (1989, p.68) argues that listening should not be limited to the oral modality, but should include sensing and observing. A good listener hears the exact words used by the interviewee (sometimes, the terminology reflects an important orientation), captures the mood and effective components, and understands the context from which the interviewee is perceiving the world.

# 6 Post 1963 Skopje earthquake reconstruction:

## Long term effects

*Vladimir B. Ladinski*

### Introduction

More than thirty years later, the 1963 Skopje earthquake still remains one of the most devastating European earthquakes of the 20th century, and one which attracted almost unprecedented levels of international co-operation and assistance. During the post earthquake period, Skopje has been transformed into a 24 km long linear city, with a population exceeding five-hundred-thousand, with wide boulevards and many new public and residential buildings. The long term effects of the internationally led, post earthquake, recovery and reconstruction are investigated in this chapter, and evaluated for the period of the three subsequent decades. The implementation of recommended development control standards in the city following the 1963 earthquake are also examined and assessed, along with the level of earthquake related site vulnerability in Skopje, based on analysis of four representative multi-family housing estates (the multi-family housing estates represent the most organised and supervised form of residential development in the city).

### Skopje's earthquake hazard

The Skopje region is one of the most seismically active areas in the Republic of Macedonia and in the Balkan, characterised by up to intensity IX MCS (Mercalli Cancani Seiberg) earthquakes, (Mihailov and Talaganov, 1985) (see Table 6.1). The first recorded earthquake dates from 518AD when 24 towns including Scupi, the capital of Dardania, were destroyed. The earthquake created a fault 45 kilometres long and 4 metres wide. The earthquake survivors moved several kilometres south-east of Scupi and created a new settlement, that developed throughout the centuries into Skopje, which was destroyed again in an earthquake in 1555. Until 1921, the seismic activity of the Skopje region was characterised by a great number of small earthquakes; whereas after the 1963 earthquake, the seismic activity of the area has been characterised by on average one intensity V MCS earthquake annually. The existence of two components, the first one determined by the local earthquakes and the other by the distant ones, is another important characteristic of the seismicity of the Skopje region, due to different dynamic properties of seismic vibrations caused by local or distant earthquakes. The lower buildings are more vulnerable in local earthquakes, while the taller are so during distance earthquakes (Mihailov and Talaganov, 1985).

**Table 6.1**
**Earthquakes with intensity VII MCS or above in Skopje**

| Location | Date | Ms | Depth (km) | Time (GMT) | Intensity (MCS) |
|---|---|---|---|---|---|
| Scupi | 518 | 6.1 | - | - | IX |
| Skopje | 1555 | - | - | - | - |
| Urosevac-Skopje | 10 Aug. 1921 | 6.2 | 20 | 14:30 | IX |
| Skopje | 7 Oct. 1921 | 4.8 | 15 | 22:24 | VII |
| Skopje | 8 Oct. 1921 | 4.8 | 15 | 02.57 | VII |
| Skopje | 26 July 1963 | 6.2 | 5 | 04:17 | IX |
| Skopje | 26 July 1963 | 4.2 | 5 | 04:32 | VII |
| Skopje | 26 July 1963 | 4.2 | 5 | 04:53 | VII |

*Source:* Based on data in Jordanovski (1993, pp.24-25); Mihailov and Talaganov (1985, p.14); Milutinovic *et al.* (1995, p.3)

*1963 Skopje earthquake*

One of the most devastating earthquakes in the area happened on 26 July 1963, when the city of Skopje was destroyed by an intensity IX MCS earthquake. Some 1,070 of the 200,000 city's inhabitants were killed, while a further 3,300 were injured, half of whom were permanently disabled (Jordanovski, 1993). Seventy-five per cent of the population were left homeless, and the earthquake destroyed or severely damaged 80.7 per cent of the total building area in the city, including 75.1 per cent of the dwelling area or 67.4 per cent of the dwellings. Eight primary and 11 secondary schools, 32 sport buildings, 9 polyclinics, 3 surgeries and pharmacies, 12 social-welfare and nurseries building were lost in the earthquake, while many other public buildings were damaged and later demolished (Milutinovic *et al.*, 1995). The earthquake's direct losses were estimated at over

$1,000 million (Jordanovski, 1993), a sum nearly equivalent to the entire annual budget of former Yugoslavia (Verney, 1979) or 15 per cent of the former Yugoslav Gross National Product for that year (Petrovski and Milutinovic, 1985; 1988), and thus exceeding the economic potential not only of the city itself, but of the whole republic (Mihailov and Talaganov, 1985). The relatively low casualty level, in comparison to the extent of destruction, can be attributed mainly to the fact that the earthquake occurred during the summer holiday season, when a number of inhabitants were away.

**Post 1963 Skopje earthquake recovery (1963-91)**

The rebuilding of Skopje is one of the most dramatic 20th century reconstruction efforts for Lagorio (1990). Immediate and long-term recovery of the city was facilitated initially with over $30 million financial aid, and resources from the federal government, the former Yugoslav republics, 82 other countries, as well as international organisation and agencies. USA, USSR, Poland, Czechoslovakia, UK, France and Germany offered in total $95.8 million reconstruction loans (but not all of these were used); 82.1 per cent of over $1,000 million rebuilding bill was met by additional federal taxes (1 per cent on income, 2 per cent on investment, etc.) 7.5 per cent was drawn from the federal reserves, and the outstanding 10.4 per cent raised by issuing government bonds (Jordanovski, 1993). According to Verney (1979), the disaster in Skopje touched the compassion of the world that led to the United Nations (UN) declaration that Skopje would be rebuilt as a gesture of international solidarity. The UN provided technical assistance by setting up the Skopje Urban Plan Project ($1.5 million) and the Vardar River Regulation Project ($1.4 million). A team of international participants, headed by Prof Adolf Ciborowski, in collaboration with the Skopje's Institute for Town Planning and Architecture (ITPA) and the City's Planning Department (CPD), developed the 1965 Skopje Urban Plan (Doxiadis Associates and Institute of Town Planning and Architecture, 1965; Polservice and Institute of Town Planning and Architecture, 1965) that guided the city's development until 1980, when it was revised (Instute of Town Planning and Architecture, 1985) in response to more rapid than expected growth. In the winter of 1962 Skopje suffered a major flood, which may have contributed to weakening of some buildings' foundations prior to the earthquake, consequently the Vardar River Regulation Project focused on regulation of the flow of the Vardar throughout the centre of Skopje to prevent future flooding. Additional UN funding of $474,200 was spent on the International Labour Organisation construction training programme (Jordanovski, 1993).

*Relief operation (July - October 1963)*

*Search and Rescue* The biggest international relief program ever mounted up until 1963 started almost immediately. Organised search, rescue and debris clearing activities started 20 to 60 minutes after the main shock, led by the Civil Protection (emergency organisation), the military garrison of Skopje (that luckily escaped the destruction of their barracks by being outside at the time) and helped by the Police force and the local population. Thanks to their efforts 6,000 of the 16,000 people that lay buried, were rescued within several hours of the earthquake (Verney, 1979). Within four hours, a complete military field hospital was operational, followed by several others by midnight on the same day and by the US Army field hospital on 28 July 1963 (Milutinovic *et al.*, 1995; Verney, 1979). The last survivor was rescued 80 hours after the earthquake (Jordanovski, 1993).

*Emergency shelter* Tents for 25,000 people were provided within 24 hours by the emergency organisation (Davis, 1975). In total 5,000 tents and 1,711 huts were supplied (Alexander, 1995). Out of 40,000 families that were housed in tents during August and September 1963, only 104 were still living in them on New Years Day 1964 (Jordanovski, 1993).

*Evacuation* After the search and rescue efforts were completed, and the homeless given shelter in tents, all those not involved in the initial reconstruction effort were evacuated (mothers with children under 18 and people over 60). Over 140,000 people left Skopje (for a period of between 3 and 9 months), leaving 60,000 in the relief force and the later rebuilding activities (Milutinovic *et al.*, 1995).

*Debris management* During the first three days, the use of heavy machinery was forbidden to prevent injury to any of the large number of people still trapped during the clearing operation. Following the three day search and rescue activities, a plan for clearing of debris and collapsed buildings was drawn up. A few days after the earthquake, damaged buildings were supported in order to prevent spontaneous collapse of the buildings from aftershocks and to salvage as much property and equipment as possible. At the same time inspection of all buildings took place, to determine whether buildings would be demolished or repaired. The initial idea of creating a specialist organisation for debris management, with participants from all construction companies was not implemented. Instead the work was carried out by local construction companies, and a 500 people strong specialised Soviet army engineering unit. Clearing activities were conducted during the following four months Some of the buildings were demolished without use of heavy machinery. This slowed the process but was economically viable as everything that could be saved was saved, with careful dismantling of the re-usable materials, (mainly plumbing, electrical and other fittings, equipment and joinery). Centralised distribution of all reclaimed materials at city level was established (Arsovski, 1985; Jordanovski, 1993; Popovic, 1990; Skartov and Stojkov, 1985).

*Burial of the dead* Over one quarter of the dead were buried in the existing cemeteries during the first day in quickly prepared graves and coffins, to prevent epidemics in the hot weather. Those that were not identified had their description and location of discovery inscribed on their graves to facilitate later identification (Jordanovski, 1993).

*Immediate housing (August 1963 - May 1964)*

Based upon the already conducted damage assessment, the authorities decided that 50,000 people should return to live in their repaired homes and 70,000 people should move into new housing. One hundred and fifty construction companies with over 25,000 workers were engaged in the construction of 14,095 new dwellings and the repair of 16,000 damaged dwellings (Jordanovski, 1993). Some 82 per cent of the new dwellings were produced in former Yugoslavia and each former Yugoslav republic financed and constructed an entire settlement for up to 15,000 people. Within eight months most of the prefabricated dwellings were in use, while the Soviet Union donated a factory for manufacturing prefabricated apartment blocks with capacity of 2,000 flats per year (Davis, 1975; Milutinovic *et al.*, 1995). During 1963-64 most of the designated dwellings were repaired, while some of them were further restrengthened in line with newly acquired knowledge in 1964-65 (Jordanovski, 1993).

Permanent housing (lasting for a period of 20 to 30 years) was built in 1963-64. The urgency and practicalities of constructing dwellings rapidly led flat sites to be selected, with dense rectilinear layouts of houses as the most convenient site plan. This was achieved by building mostly timber frame prefabricated houses on sites, away from the damaged city centre and on agricultural land. The effort and resources put into the immediate housing might have been at the expense of the final reconstruction. The creation of 18 new estates away from the damaged residential areas drew resources and focused the activity away from the old settlement, contributing to further delay in reconstruction activities. In addition, there is the alien nature of the newly built prefabricated houses (being culturally unfamiliar at a time when, psychologically, the community is trying to recover normality and needs familiarity, (Coburn and Spence, 1992)); along with the disrupted local social structures, the broken neighbourhood cohesiveness and the fear of fire which were also some of the negative effects on the community in its recovery from the earthquake.

In contrast after the 1976 Great Tangshan earthquake, the immediate housing was provided by building temporary shacks and rebuilding residential units, in the same areas, using local materials and masonry rubble from the devastated buildings. These temporary dwellings allowed whole families to live without the need for another move until new permanent housing was completed (Lagorio,

1990; Yong *et al.*, 1988). This policy preserved local social structures and allowed appropriate time for careful reconstruction planning (Lagorio, 1990). Similarly in Mexico City, temporary housing of heavy craft board and corrugated sheet metal was assembled, in streets closed to traffic, adjacent to construction sites designated for the permanent housing program that followed. This allowed the displaced residents to maintain neighbourhood cohesiveness and actively participate in the rehabilitation and repair of their damaged homes or the construction of new residential units (Lagorio, 1990). Unfortunately the social structures and networks in Skopje were heavily disrupted, by the implemented recovery policies. Furthermore, the new residential units were constructed without residents' participation, due to the level of destruction as well as government policies; since countries with central governments, as was the case in former Yugoslavia, usually did not encourage residents' participation in the decision process (Lagorio, 1990).

*Replanning (1963-66)*

The federal government considered all planning options for reconstruction including complete relocation of the city to a new site. However, after many studies of soil conditions and seismic vulnerability of the region were conducted, it was finally decided that no alternative site existed in the republic that would be more appropriate for the capital city; since all other alternative sites considered were equally vulnerable (see Figure 6.1). A decision was therefore made to rebuild the city on the same site, but with steps included in the reconstruction program to mitigate seismic hazards.

The main rationale behind the decision was that, in economic terms, although most of the population was homeless, 80 per cent of Skopje's social and financial capital (in transportation facilities, public utility services, operational factories, repairable building stock, and undamaged structures) was still in place (Arsovski, 1989; Lagorio, 1990). Most of the factories were in working order and they represented the major concentration of industrial employment in Macedonia. In 1962, 35 per cent of the entire industry of the republic was concentrated in Skopje and was responsible for 43 per cent of the entire income of the republic (Jordanovski, 1993). Attention was therefore, focused on rebuilding of the disproportionately destroyed housing. The period up to December 1964 concentrated on the restoration of essential services, repair of damaged buildings, getting factories back into full production and caring for all homeless families; while the period between 1965 and 1971 concentrated on the planned reconstruction which followed the 1965 Plan, influenced by seismic factors (Davis, 1975).

The first stages of reconstruction and redevelopment (3 to 5 years) focused outside the city centre. In addition, a federal law gave the city the power to acquire, (at current market value), any piece of property within the urban area for its urban plan (Greene, 1987; Lagorio, 1990). The officials wanted to develop a separate, but complementary plan for the damaged city centre. The postponement of the reconstruction of the city centre allowed time for soil analysis and studies of local seismic conditions, and an international competition for the City Centre Master Plan (Master Plan).

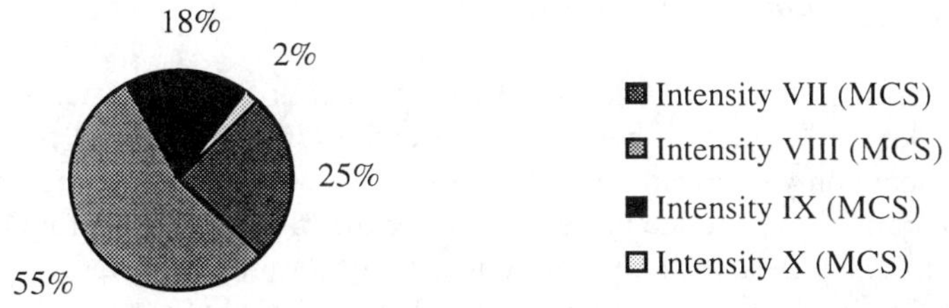


**Figure 6.1 Territorial distribution of expected earthquakes in the Republic of Macedonia**

*Source:* Based on data in Jordanovski (1993, pp.23-24)

This Master Plan competition, the first one ever (for a post earthquake reconstruction project), was an international design competition, held in 1965, for a Master plan of the city centre and sponsored by the UN and federal government. The two winning plans were totally different in character. Kenzo Tange's proposal (60 per cent joint winner) was a concept of an inner core surrounded by a City Wall of residential accommodation and a City Gate as a traffic and administrative centre (Davis, 1975; Kultermann, 1970). Some found Tange's solution impractical, unrealistic, and out of proportion to the available limited resources (Lagorio, 1990). However others preferred the transportation linkage and social communication systems, provision of many open spaces, and the overall decentralised concept, which mitigated some of the deficiencies of the congested, more vulnerable parts of the old city. The 40 per cent joint winner, the Croatian Town Planning Institute in Zagreb, produced a more modest flexible proposal to the changing needs (Davis, 1975). A compromise, therefore, was established. Vojislav Mackic was nominated to lead a team formed from the joint winners, with responsibility for the final design and its implementation. By July 1966, the team completed their work on the Master Plan, influenced mostly by Tange's proposals (Lagorio, 1990).

The results of the seismicity studies of the Skopje region were incorporated into the planning process, by identifying the safe development areas and those where further soil analysis would be required. The general restrictions for the future city development were also established (Greene, 1987; Lagorio, 1990). The federal 1964 Code for building design and construction in seismic areas was adopted, in addition to the recommendations incorporated in the 1965 Plan to mitigate further earthquake hazard. Furthermore, the City's new Fire Safety Code was adopted in the early 1970s.

*Reconstruction (1965-71)*

According to Jordanovski (1993), 98 per cent of the financial resources allocated for the reconstruction between 1963 and 1970 were used in the period 1963-73. The economy received reconstruction funds 5.3 fold higher than the actual losses, apparently, to generate more income for the bigger and better modern city, as well as to take advantage of the opportunity to improve both the city's and the republic's economy (Jordanovski, 1993). The reconstruction investment in the industry resulted in a 74 per cent employment increase in the 10 year period following the disaster. Agriculture also gained, with three fold higher production, but lost 1,000 hectares of land in the Skopje valley for the reconstruction activities. The construction industry had a 70 per cent rise in employment, and from 1966 took over most of the reconstruction activities. City Shopping Centre, Skopje Fair, five superstores, Telecommunication Centre, the Main Post Office, Skopje Airport, hotels Panorama, Belvi, Kontinental, Skopje and Olimpisko Selo etc., were all built in that period and many other building and complexes were reconstructed.

The construction of some public projects was deliberately delayed. The Skopje's Fund for Reconstruction and Development that managed the post earthquake reconstruction financial resources, decided to finance some of the projects in the period 1971-80 from their annual income, such as Regulation of the Vardar river, the Transportation Centre, some major city roads, the Cultural Centre, the Radio and Television Centre, reconstruction of the Old Market, the Municipal Court, etc. Although most of the construction activities started, nearly all mentioned projects were not fully completed due to the lack of financial resources. The income generated from the Fund was not sufficient to sustain construction requirements and the race against inflation at the same time.

*Post earthquake development trends (1963-91)*

*Population* By 1971 the population of Skopje has risen 16 per cent above the 1965 estimate, while in 1981 the population growth was underestimated by 17 per cent (see Figure 6.2). Furthermore, these growth trends were extended on the basis of the relationship between the city and the region, and especially with the republic (see Figures 6.3 and 6.4). The population of Skopje represented 14 per cent of the population of the republic (1961) and increased to 22 per cent (1991). The population of Skopje Region was 19 per cent of the total population of the republic in 1961, while in 1991 it was 28 per cent. The 1991 census indicated that the 1985 Amendments have since overestimated the population growth in Skopje by 11 per cent.

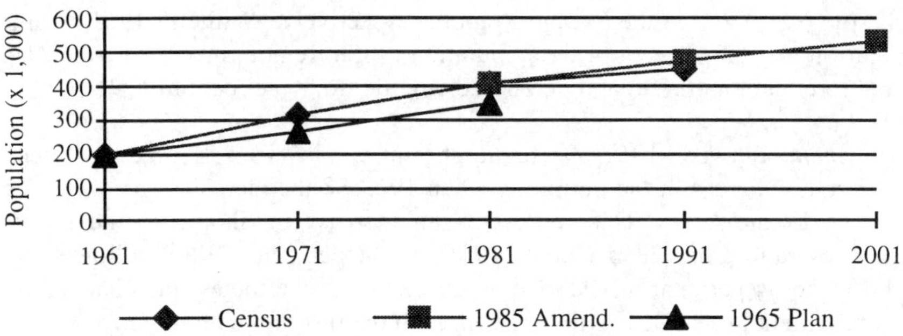


**Figure 6.2 Population trends in Skopje (City)**

*Source:* Based on the 1961, 1971, 1981 and 1991 Census and planning documents predictions (1965 Skopje Urban Plan: for the period 1961-81 and 1985 Amendments to the Skopje Urban Plan: for the period 1981-2001) (Institute on Town Planning and Architecture, 1985)

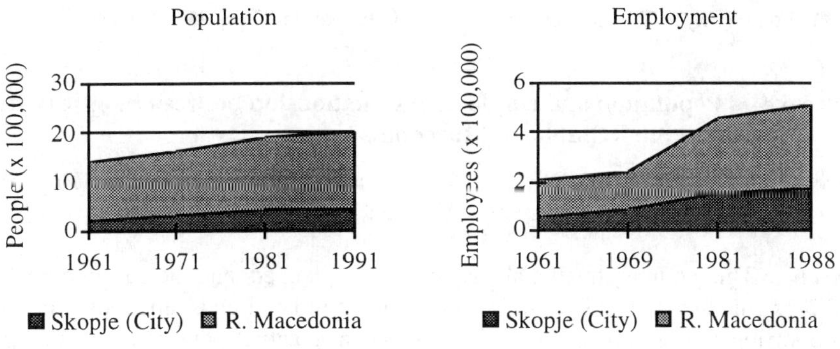


**Figure 6.3 Population and employment in Skopje (City) and the Republic of Macedonia**

*Source:* Based on the 1961, 1971, 1981 and 1991 Census and the Republic of Macedonia Statistical Office data

The relationship between the population of Skopje and the republic (see Figures 6.3 and 6.4) is of great importance from an earthquake protection standpoint, since the population concentration is one of the main vulnerability factors. The greatest increase was recorded between 1961-81, and now, more than 20 per cent and 25 per cent of the population in the Republic of Macedonia are concentrated in Skopje (City) and in the Skopje Region, respectively. While in 1961 Skopje's population was 73 per cent of the population of the Region, in 1981 it was 81 per cent (11 per cent increase) and decreased slightly to 80 per cent in 1991.

*Employment* Between 1961-69, the number of employees in Skopje has risen by 62 per cent, whereas in the period between 1969-88 the increase was only 52 per cent (see Figure 6.3). This represents an increase in Skopje to the republic employees ratio from 28 per cent (1961) to 38 per cent (1969) followed by a slight fall to 34 per cent in 1988 (see Figure 6.4). Nevertheless the concentration of employment in Skopje is still higher than in the rest of the republic.

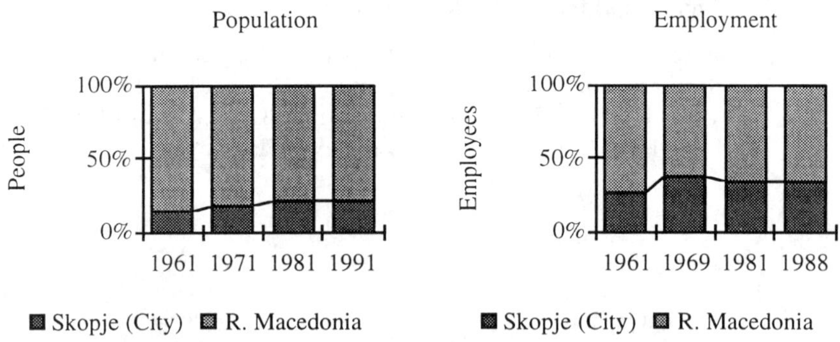


**Figure 6.4** **Population and employment relationship between Skopje (City) and the Republic of Macedonia**

*Source:* Based on the 1961, 1971, 1981 and 1991 Census and the Republic of Macedonia Statistical Office data

*Housing* The comparative analysis, between planned and actual number of dwellings in the city of Skopje, indicates similar trends to those of the disproportion that exists between the planned and actual population in the city (see Figures 6.2 and 6.5). The 1965 Plan underestimated the construction of dwellings by 19 per cent in 1971 and 4 per cent in 1981, respectively, while the 1985 Amendments overestimated the dwellings construction by five per cent (1991).

During the period 1966-71, 26,700 new dwelling units were built. As a result, eight years after the disaster, with the prefabricated, reconstructed and the housing which survived, Skopje had in total 74,000 dwellings in comparison with 36,578 dwellings in 1963. Between 1982-91 only 87 per cent of the planned dwellings were constructed. At the same time, the proportion between the public and private dwelling construction varied (e.g. the public dwelling construction figures were 83 per cent in 1982 compared to only 24 per cent in 1987) (see Figure 6.6). Fifty-three per cent of all dwellings built in the observed period were public, while the rest were private. The average floor area of dwellings built between 1982-91 was 67m$^2$ (70m$^2$ in the public sector and 63m$^2$ in the private). Also, there is a difference in the dwellings' structure (see Figure 6.6), 45 per cent of all dwellings constructed in Skopje between 1982-91 in the public sector were one bedroom; while in the private sector there was almost even distribution among the construction of studio, 1 and 2 bedroom properties. The construction of four and more bedrooms dwellings was nearly four fold higher in the private sector than in the public.

A substantial increase in housing standards has been recorded in the past years. The average housing floor area per person in the city increased from 8.50m$^2$ (1963) and 10.60m$^2$ (1965) to 14.14m$^2$ (1981) and 15.00m$^2$ (1991).

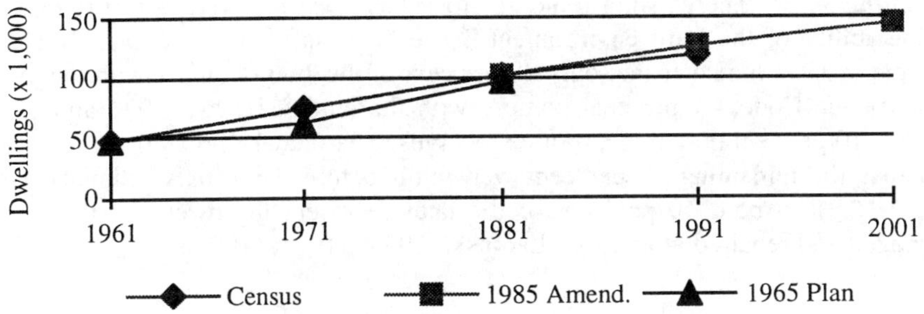


**Figure 6.5** **Housing trends in Skopje**

*Source:* Based on the 1961, 1971, 1981 and 1991 Census and the Republic of Macedonia Statistical Office data

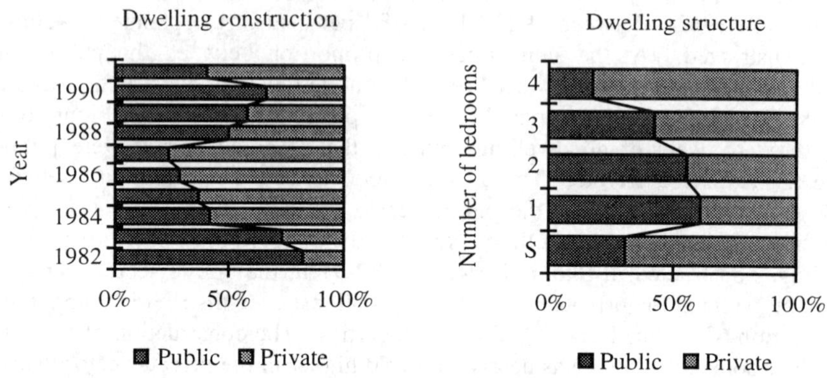


**Figure 6.6 The relationship between public and private dwellings construction in Skopje**

*Source*: Based on the Republic of Macedonia Statistical Office data

*Dwelling stock* The existing building stock has great importance to the overall vulnerability of the built environment. The analysis of the dwellings stock in Skopje in 1991 indicates that only 20 per cent of dwellings comply with the 1981 Antiseismic Code, 41 per cent comply with the outdated 1964, 1971 and 1977 Codes, 10 per cent do not incorporate recognised earthquake protection measures, whereas the remaining 29 per cent were built before the 1963 earthquake (see Figure 6.7). About 40 per cent of the later represent the dwellings that were damaged and repaired after 1963 (Ladinski, 1994b, 1994c, 1995b).

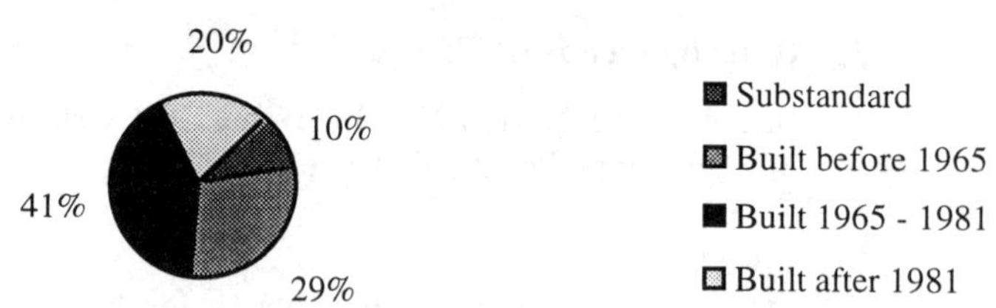


**Figure 6.7 Existing dwelling stock in Skopje in 1991**

*Source*: Ladinski (1994b)

*Post earthquake housing estates*

The implementation of the recommended development standards and earthquake protection measures can be followed most easily on the post-earthquake multi-family housing estates, as they were the most organised and supervised form of development in Skopje. Based on previously identified estate layout types in Skopje (Vlcevski, 1984), the following multi-family housing estates have been analysed:

- Karpos IV (K-IV) with freely located residential blocks developed in the immediate post earthquake period with 1980s additions (see Figure 6.8);
- City Wall (CW) with system of open residential blocks creating a street patterns developed in 1970s and 1980s, as a part of the City Centre Master Plan (see Figure 6.9);
- Kapistec (K-II) with freely concentrated groups of residential blocks in development since 1970s (see Figure 6.10); and
- Aerodrom I (A-I) with concentrated groups of residential blocks creating pedestrian streets, mostly developed in 1980s (see Figure 6.11).

The last three estates are analysed through their representative and repetitive neighbourhood units (City Wall T1-8/B3-6, Kapistec II and Aerodrom A1-1).

*Development control standards* The layout plans of the estates were used for calculation of their development indicators (see Table 6.2). Where appropriate the development indicators were calculated as net (residential areas only) and gross (residential and non-residential areas within the estate). Various recommended development control standards for earthquake prone areas were identified (see Table 6.3). They were used to compare the actual development level of the estates with the recommended one.

The comparison between the analysed development indicators for each estate (with the recommended development control standards in earthquake prone areas) clearly shows that most of them have been exceeded in the central City Wall and Kapistec estates. The gross residential density is 91 per cent higher in the City Wall and 33 per cent in Kapistec, while the net residential density is 83 per cent higher in Kapistec and 42 per cent in the City Wall than the recommended (see Figure 6.12). The net floor space index of all estates is well above the recommended, ranging from 2- (Karpos IV) to up to 5- fold higher in the City Wall and Kapistec (see Figure 6.13). The net building plot ratio of all estates is within the recommended, with the exception of the City Wall where it is 16 per cent higher (see Figure 6.12). The gross floor space index and the gross building plot ratio are similar to their net relationships.

The combination of average net building plot ratio and higher net floor space index indicates the main cause of the problems - very tall buildings for the available sites. Even before the construction of the Kapistec estate started, Galic

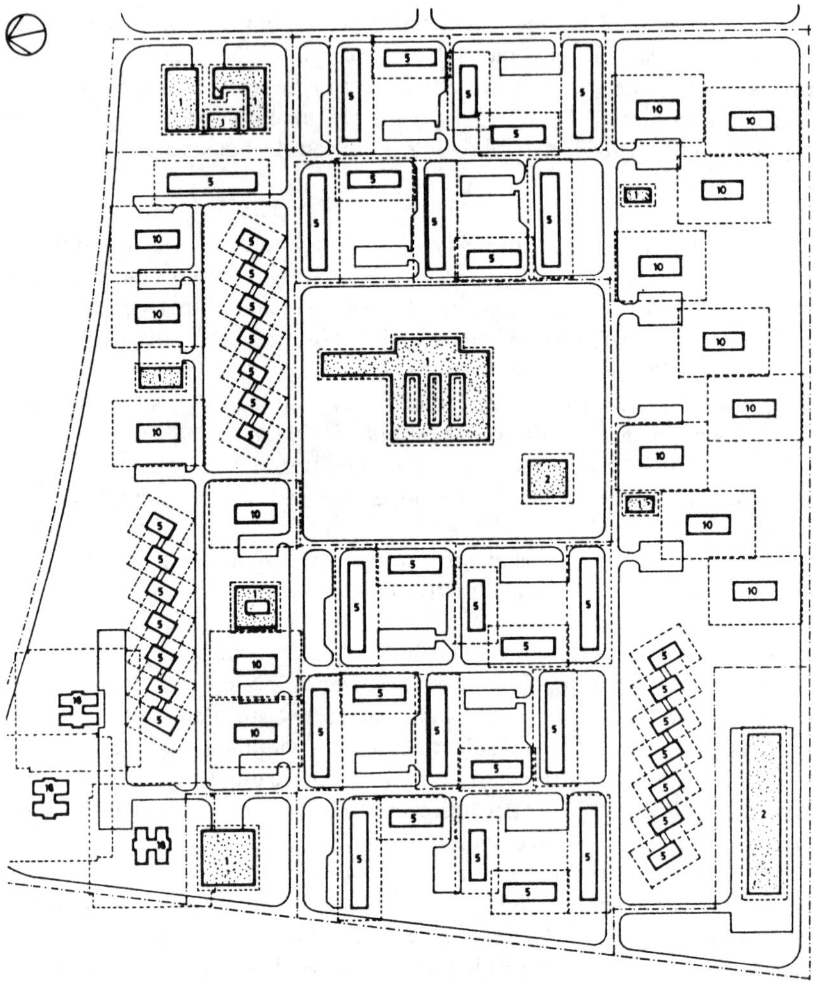


No. number of storeys; - - -  debris extent line; -.-.-  boundary line;
:::  non-residential building

**Figure 6.8  Karpos IV (K-IV) debris layout plan (1:5000)**

*Source*: Ladinski (1994b)

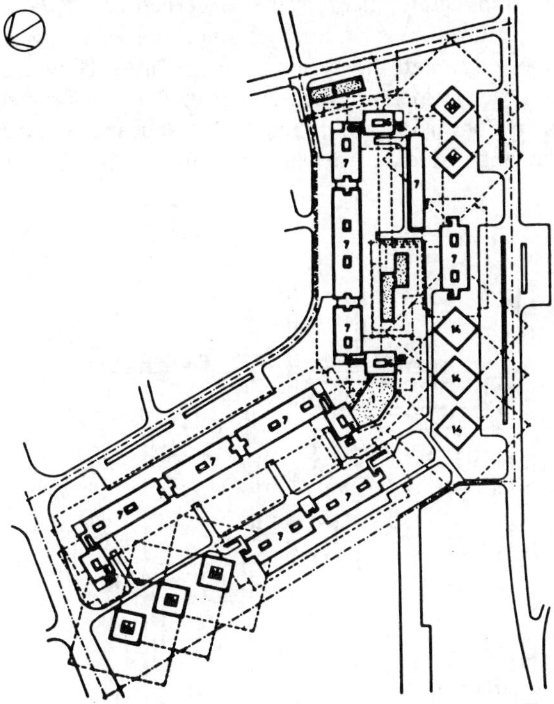


No. number of storeys; - - -   debris extent line; -.-.-   boundary line;
:::   non-residential building

**Figure 6.9   City Wall (CW) debris layout plan (1:5000)**

*Source*:   Ladinski (1994b)

(1977) pointed out that the estate was based on false economic effectiveness, since with the increase of the densities above 300 bed spaces per hectare, construction costs would increase and quality of life would decrease. Others (Skopje Faculty of Architecture, 1977) were concerned about the environmental implications on the central areas of Skopje, since the estate was effectively a barrier between the city and the Vodno mountain. Their concerns were disregarded and, in order to increase the number of flats constructed on some of the buildings, floor to floor heights were reduced to squeeze an extra storey on to the buildings. The net residential density in Kapistec is 735 bed spaces per

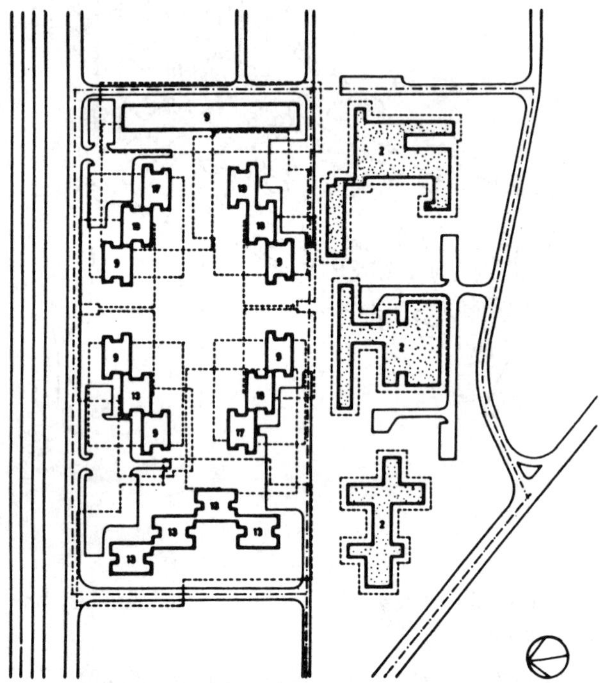

No. number of storeys; - - - debris extent line; -.-.- boundary line;
::: non-residential building

**Figure 6.10 Kapistec II (K-II) debris layout plan (1:5000)**

*Source*: Ladinski (1994b)

hectare, and it does not take into account the real occupancy. Furthermore, earlier studies (Blazevska, 1982, 1983) indicate that up to 30 per cent of the flats on the estate were inhabited by more residents than planned.

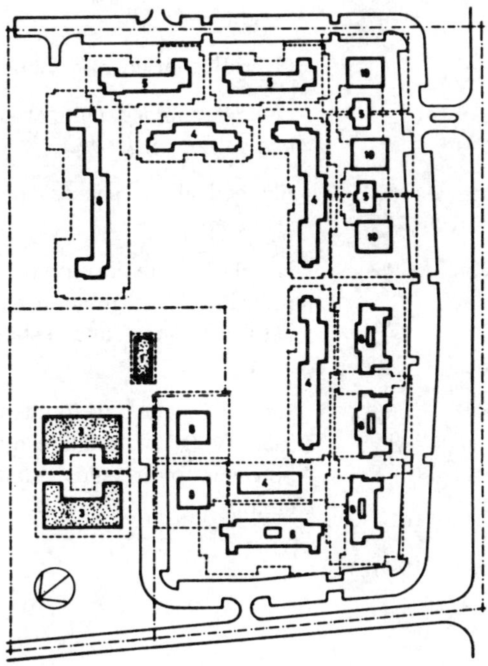

No. number of storeys; - - - debris extent line; -.-.- boundary line;
::: non-residential building

**Figure 6.11 Aerodrom I (A-I) debris layout plan (1:5000)**

*Source*: Ladinski (1994b)

**Table 6.2**
**Parameters used in the development control analysis**

| PARAMETER | FORMULA | NOTE | UNIT |
|---|---|---|---|
| Estate area ($A_e$) | - | Overall estate area | ha |
| Built area ($A_b$) | - | Area of estate under buildings | ha |
| Floor area ($A_f$) | - | Overall floor area on estate | ha |
| Bed spaces ($b_s$) | - | Overall planned number of bed spaces on estate | - |
| Residential density ($D_r$) | $\frac{b_s}{A_e}$ | Planned bed spaces per estate area | $b_s$/ha |
| Floor space index ($I_{fs}$) | $\frac{A_f}{A_e}$ | Level of estate development indicating relationship between floor and estate area | - |
| Building plot ratio ($R_{bp}$) | $\frac{100 A_b}{A_e}$ | Level of estate development identifying relationship between built and estate area | per cent |

*Source*: Ladinski (1994b)

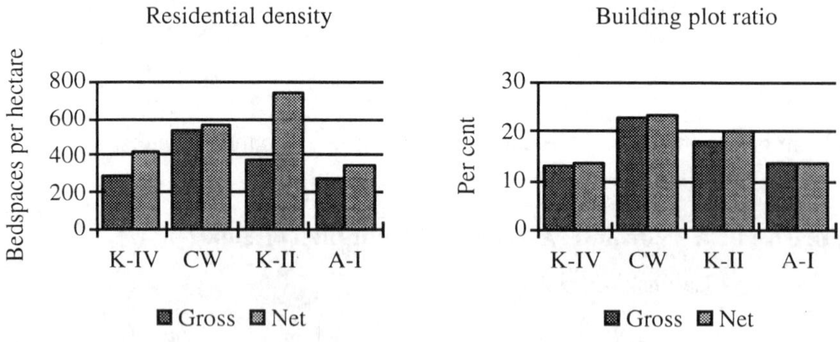


**Figure 6.12 Residential density and building plot ratio**

*Source*: Ladinski (1994b)

**Table 6.3**
**Recommended development control standards for earthquake prone areas in former Yugoslavia**

| Development Indicators | 1965 Skopje Urban Plan | Zakic (1982) | Markov (1982 and 1983) | Vlcevski (1984) | 1985 Amendments | Krstic (1987) |
|---|---|---|---|---|---|---|
| ${}_{g}D_{r}$ ($b_s$/ha) | - | - | - | 250-*280* | - | <350[a] |
| ${}_{n}D_{r}$ ($b_s$/ha) | *<400* | - | - | < *400* | <300 | <500 |
| ${}_{g}I_{fs}$ | - | - | <2.00<br>(<0.33)[b] | - | 0.65-*0.75* | - |
| ${}_{n}I_{fs}$ | - | *1* - 1.20 | *<1*<br>(<0.33)[b] | - | 0.36-*0.42* | - |
| ${}_{g}B_{pr}$ (%) | - | - | <40<br>*<20*[b] | - | - | <40 |
| ${}_{n}B_{pr}$ (%) | - | *20*-30 | <30<br>*<20*[b] | - | - | <30 |

prefixes: $_{g}$ - gross, $_{n}$ - net

[a] - for towns with population of more than 100,000
[b] - for prevention of follow on earthquake hazards (e.g., fire)
NOTE: figures in italics used for comparison with actual results.

*Source*: Ladinski (1994b)

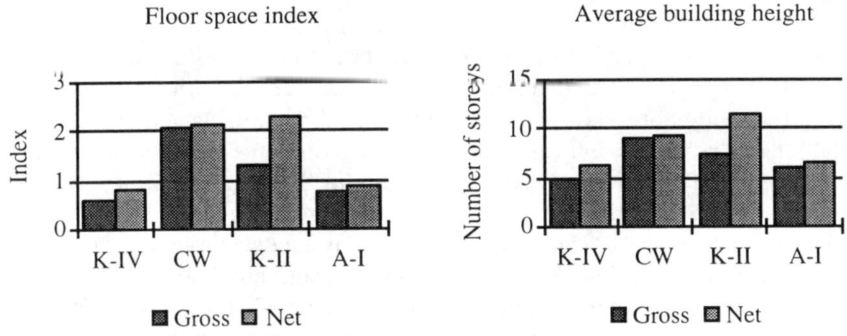


**Figure 6.13 Floor space index and average building height**

*Source*: Ladinski (1994b)

*Site vulnerability* The site vulnerability assessment of the estates was conducted using the methodology developed in the 1970s (Kovraz, 1985; Markov, 1982, 1983; Zakic, 1982, 1983). This technique assesses the maximum potential vulnerability of the area among the buildings on a site, as a result of natural or man-made disasters, and assumes full collapse of all buildings on the site. The hypothetically created debris on site leads to an increase of potential site vulnerability as debris obstructs relief activities. It is assumed that a building collapses symmetrically to both axes, with extension of the debris (created by the collapse) of one half of the building height (measured from the building perimeter) and maximum debris height of 20 per cent of the building height (measured at the building perimeter) that gradually decreases to zero at the maximum debris extend line. Thus the site vulnerability analysis is based mostly on debris quanitiy, and building collapse geometry estimates (see Table 6.4) derived from analytical, statistcal and experimental data. The building collapse geometry is expressed through the debirs layout plan of the site (indicating symmetrical collapse of all builds on the site with indicted extent of the debris) and the relationship between a building and a debris height (Kovraz, 1985).

Debris layout plans can be used to evaluate the potential level of site coverage with debris, the provision for access-egress and the possible location of refuge space, where residents can be temporarily evacuated to protect them from follow on earthquake hazards, such as fire, or further collapses of damaged buildings. The coverage of observed area with debris (debris plot ratio) can be used to estimate the vulnerability of that area and is a very useful tool for layout planning and selection of appropriate building heights. Building and debris height, as well as debris quantity, can also be used to estimate the site vulnerability (Kovraz, 1985).

In order to allow further assessment of the consequences of debris allocation on a site, the debris overlapping ratio parameter was introduced (Ladinski, 1994a, 1994b, 1995a). This ratio represents the level of debris overlapping among all buildings on the site. It points out the existence of areas on a site where debris concentration might occur. These areas of debris concentration can be identified on the debris layout plan and they are an important indicator for identification of spots where access-egress facilities might be reduced and where recovery and relief efforts might be delayed due to additional debris clearing activities.

The site vulnerability indicators were calculated on the basis of debris layout plans for each of the analysed estates in the city of Skopje. The results were used to compare the actual site vulnerability indicators with the recommended. According to Kovraz (1985), up to 40 per cent is acceptable level of debris:plot ratio and up to four storeys is the acceptable average building height, while the potential existence of debris overlapping should be avoided (Ladinski, 1994b).

The comparison between the actual analysed site vulnerability indicators for each estate with those recommended, clearly indicates that they have been exceeded on all the estates. The net debris:plot ratio is higher in all estates, from 21 per cent in Aerodrom I to over 200 per cent in Kapistec (see Figure 6.14). The

**Table 6.4**
**Selected site vulnerability parameters**

| PARAMETER | FORMULA | NOTE | UNIT |
|---|---|---|---|
| Average building height *($H_{ab}$)* | $\frac{A_f}{A_b}$ | Average building height of estate | storeys |
| Building height *($H$)* | - | Building height above ground | m |
| Building debris height *($h_d$)* | $h_d = 0.2\ H$ | Assumed average debris height along perimeter of fully collapsed building | m |
| Building debris extent *($d_e$)* | $d_e = 0.5\ H$ | Assumed debris extent from fully collapsed building | m |
| Building debris area *($a_d$)* | - | Area covered by debris from fully collapsed building | ha |
| Expected debris area *($A_{ed}$)* | $\Sigma\ a_d$ | Sum of all individual building debris areas $a_d$ on estate regardless of their relationship | ha |
| Debris area *($A_d$)* | - | Area covered by debris from all buildings fully collapsed on estate | ha |
| Debris plot ratio *($R_{dp}$)* | $\frac{100\ A_d}{A_e}$ | Level of estate coverage with debris identifying relationship between debris area and estate area | % |
| Debris overlapping ratio *($R_{do}$)* | $\frac{100\ (A_{ed} - A_d)}{A_{ed}}$ | Level of debris overlapping on estate identifying relationship between the potentially overall expected and assumed actual debris area | % |

*Source:* Based on Kovraz (1985); Ladinski (1994a, 1994b, 1995a)

net debris overlapping ratio is between 8 per cent (Aerodrom I) and 37 per cent (Kapistec), indicating the existence of areas of debris concentration on all the analysed estates (see Figure 6.14). Also, the average net building height on all estates is between 1.5- (Karpos IV) and 3- fold (Kapistec) higher than that recommended (Figure 6.13). The gross debris plot ratio, the gross debris overlapping ratio and the gross average building height are similar to their net relationships.

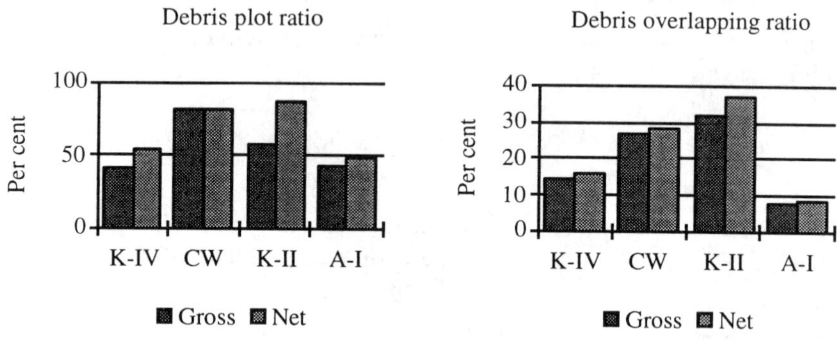


**Figure 6.14 Debris plot and overlapping ratios**

*Source*: Ladinski (1994b)

The debris layout plans indicate the maximum expected debris extent and debris overlapping on the analysed estates. In Karpos IV, there is a concentration of areas of potential debris overlapping in the north-west corner of the estate, where the 18 storeys tower blocks have been added after the initial development was completed. In addition, the other significant areas of debris overlapping are identified around the 3 groups of 7 interconnected 5 storey blocks (see Figure 6.8). The City Wall estate is characterised by potential debris overlapping among all the adjacent buildings (see Figure 6.9). In case of a devastating earthquake it would be very difficult to access or evacuate the primary school, located in the centre of the development and surrounded by potentially overlapped debris. In Kapistec, there are two completely different areas. While at the north (residential) part of the estate potential debris overlapping is to be expected almost everywhere, the south (non-residential) part of the estate is potentially free of debris overlapping (see Figure 6.10). Aerodrom I is characterised by almost evenly distributed potential debris overlapping, although it is lower than that in Kapistec or the City Wall estates (see Figure 6.11). Despite being smaller in its extent, the location of the potential debris overlapping will reduce access and evacuation in case of a disaster, since the building entrances are from the inner pedestrian street between the buildings The provision of desired 3 $m^2/b_s$ refuge space is partially fulfilled only at Kapistec (2.72 $m^2/b_s$) and Aerodrom I (2.57 $m^2/b_s$) while there is no provision at all in the City Wall and only 1 $m^2/b_s$ in Karpos IV (see Figure 6.15).

The site vulnerability analysis confirms the findings of the development control standard implementation assessment. The average net building height of all estates is above the recommended, as well as the net debris plot ratio. The net debris overlapping ratio ranging from 8 per cent (Aerodrom I) to up 37 per cent (Kapistec) indicates that there are zones on the estates with only limited access-egress. From site vulnerability point (excluding the soil conditions and the buildings own vulnerability influences), the safest estates seem to be Aerodrom I and Karpos IV. The research did not consider the environmental issues, since Aerodrom I is surrounded by steel plant and adjacent garbage disposal site on north, chemical complex on east, and cement and glass factories on south.

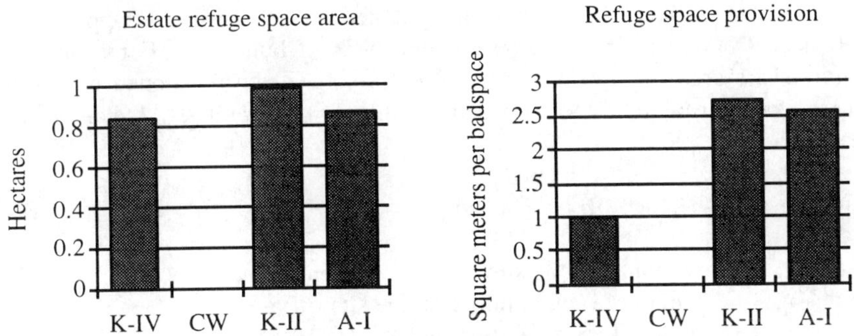


**Figure 6.15 Gross estate refuge space area and refuge space provision**

*Source*: Ladinski (1994b)

*Distance between buildings* The appropriate distance between buildings is one of the oldest methods of improving safety in earthquake areas and is important for provision of access-egress, prevention of fire spreading after disasters, and for adequate provision of sunlight (especially in residential areas). A number of recommended criteria were identified and used for comparison with actual distances among buildings on the observed estates (see Table 6.5).

The analysis of distance among buildings on the analysed estates, based on their layout plans, indicates that in most cases the distances between buildings are more generous among the low rise buildings than in the high rise. The buildings in the older part of the Karpos IV estates are further apart than the newer group of three tower blocks. The 14 storey tower blocks of the City Wall estate are less than 10 meters apart at their closest points. In Kapistec the distance between lower buildings is greater than between higher, while in Aerodrom I, the distances among buildings are kept more or less identical regardless of different building heights.

The distance between buildings varies among and within the estates. In the City Wall and Kapistec estates the distances between the lower buildings are greater than between the taller buildings. Thus, they increase the site vulnerability among the buildings with more residents. In Aerodrom I, the almost identical distance among the buildings does not take account of the difference in their height. The location of the primary school in the middle of the City Wall estate raises concerns regarding the potential exposure of the pupils to involuntary risk, since in a case of a disaster there is a potential situation of limited access-egress and evacuation provision from the school. In addition, most of the taller buildings are located on the south side of the estates. Thus they interfere with the sunlight requirements.

Based on the required level of protection, it is possible to determine the required buildings heights for known site area and buildings dimensions and vice versa, as well as layout of the buildings in such a manner that no debris overlapping occurs and access-egress is provided on all four sides of buildings. For example the required site area for a estate where fire spread will be postponed and limited access-egress provided (recommended distance $l = 0.5\,H_1 + 6 + 0.5\,H_2$ ) should be:

$$A_e \geq \Sigma\,(a_n + H_n + 6)\,(b_n + H_n + 6)$$

where $A_e$ = total area of the site for development ($m^2$),
$a_n$ = length of each building on the site (m),
$b_n$ = width of each building on the site (m)
$H_n$ = height of each building on the site (m).

## Reconstruction problems (1963-91)

### *Post earthquake growth*

Probably the greatest post earthquake problem in Skopje has been its growth rate. The reconstruction-generated activities led Skopje to become a boomtown and thriving regional centre, initiating a very rapid growth. For many years, especially during the late years of the 1960s recession, Skopje was the biggest construction site in former Yugoslavia, attracting thousands in search of better jobs and housing. The city also provided jobs and hopes for a better life for the people in surrounding villages and countryside. Therefore the authorities attempted to discourage migration into the city, by restricting the sale of public dwellings to people with at least five years residence and employment in the city and by locating industry in surrounding cities. Regardless of these efforts, the authorities were only partially successful in curbing the migration from 7 per cent during 1963-73 to between 3 and 4 per cent in the mid 1980s (Greene, 1987).

It seems that the decentralisation policy introduced with the 1985 Amendments produced some positive results. The proportion of Skopje's population in the overall population of the republic increased from 19 per cent (1961) to 26 per cent (1981), followed by more sustainable increase to 28 per cent (1991). The

**Table 6.5**
**Various recommended distances between buildings**

| SOURCE | DISTANCE BETWEEN BUILDINGS $l$ (m) | NOTE |
|---|---|---|
| Doxiadis Assoc. and ITPA (1965) | $l = 2.3\ H_1$ $(H_1 \geq H_2)$ | Provides 2 hours sunlight per day for south faced buildings for latitude 41°50' (Skopje) |
| Doxiadis Assoc. and ITPA (1965) | $l = 2\ H_1$ $(H_1 \geq H_2)$ | Provides about 2 hours sunlight per day for south faced buildings for latitude 41°50' (Skopje) |
| Doxiadis Assoc. and ITPA (1965) | $l = 2\ H_1$ $(H_1 \geq H_2)$ | Provides 6 hours sunlight per day for east-west faced buildings for latitude 41°50' (Skopje) |
| Doxiadis Assoc. and ITPA (1965) | $l = H_1$ $(H_1 \geq H_2)$ | Provides 2 hours sunlight per day for east-west faced buildings for latitude 41°50' (Skopje) |
| Kovraz (1985) | $l = H_1 + 20 + H_2$ | Prevents fire spread and provides full access-egress |
| Kovraz (1985) | $l = H_1 + 10 + H_2$ | Prevents fire spread and provides nearly full access-egress |
| Kovraz (1985) | $l = H_1 + H_2$ | Reduces fire spread and provides conditional access-egress |
| Kovraz (1985) | $l = 0.5\ H_1 + 6 + 0.5\ H_2$ | Postpones fire spread and provides limited access-egress |
| Kovraz (1985) | $l = H_1 + 3 + H_2$ | Provides some access-egress in case of debris extent: $d_e = H$ |
| Kovraz (1985) | $l = 0.5\ H_1 + 3 + 0.5\ H_2$ | Provides some access-egress in case of debris extent: $d_e = 0.5\ H$ |
| Krstic (1987) | $l = 0.5\ H_1 + 5 + 0.5\ H_2$ $l \geq H_1\ (H_1 \geq H_2)$ | Recommended distance between buildings in earthquake areas in Croatia |
| Kovacev (1987) | $l = 0.5\ H_1 + 0.5\ H_2$ | Recommended distance between buildings in earthquake areas |
| Popovic (1990) | $l = 1.5\ H_1$ $(H_1 \geq H_2)$ | Recommended distance between buildings in earthquake areas |
| Lagorio (1990) | $l = 0.5\ H_1 + 0.5\ H_2$ | Recommended building height in Mexico City after the 1985 earthquake |

*Source*: Ladinski (1994b).

concentration of employment in Skopje shows a downward trend since early 1970s, coinciding with the ending of the reconstruction funds, and especially during 1980s. Nevertheless, 34 per cent (1988) of the overall employees in the Republic of Macedonia and 28 per cent (1991) of the overall population of the republic are concentrated in Skopje. These figures give serious reasons for concern, since nearly a third of the population and of the employment, are concentrated in a single city located on an expected intensity IX MCS earthquake prone area.

Skopje has to try to curb its growth in order to reduce future earthquake vulnerability and increase seismic safety. After the 1963 earthquake, the densities were reduced. According to the 1965 Plan, the city's urban zone was enlarged three times to accommodate a similar number of people and activities as before the disaster. Since 1980s, these same 'reduced densities' are being adjusted to accommodate the rapid growth by adding storeys to proposed or existing buildings and by building smaller flats in them. Thus, the more rapid the growth the more difficult it is to comply with the original intent of low densities to minimise seismic vulnerability (Greene, 1987).

*Substandard and illegal housing*

The analysis of the ten settlements with the lowest standards in Skopje in 1981 (Vlcevski, 1984), indicated that most of these settlements were without or had very little urban equipment and infrastructure, but did have a number of illegally constructed buildings (see Table 6.6). They were also characterised by a high proportion of children aged 0-10 living there (38.05 to 48.64 per cent in Senic and Suto Orizari, respectively), in overcrowded dwellings with between 6.2m$^2$ (N. Karev) and 11.62m$^2$ (Keramidnica) floor area per occupant (see Figure 6.16). In addition, between 63.20 per cent (Lirija) and 100 per cent (Keramidnica) of the buildings in these settlements were constructed using substandard materials (see Figure 6.16). Although some improvements have been made since, the conditions are not noticeably improved, and these areas remain potentially more vulnerable to future disaster.

The state of the substandard areas indicates the inability of the authorities to control single family housing construction in Skopje as being another aspect of the reconstruction problems, which is a direct result of the rapid growth, coupled with the housing demand and policies. The construction of some small (often self built) buildings took place, without the required permits, resulting in buildings built with substandard materials or construction methods. This type of development, called Wild Development, occurs in courtyards of existing houses or on the city's fringes, very often on agricultural land or green belt areas. According to Greene (1987), 12 per cent of the population of Skopje lived in such dwellings in 1981. 10,480 and 20,555 Wild Development dwellings were built between 1976 and 1982 in Skopje and in the Republic of Macedonia, respectively (Bobic and Vujovic, 1985). Fifty-one per cent of the overall Wild Development

**Table 6.6**
**Settlements with lowest standards in Skopje in 1981**

| SETTLEMENT | NUMBER OF ILLEGAL BUILDINGS | URBAN EQUIPMENT |
|---|---|---|
| Keramidnica (K) | >100 | partial |
| V. Karangeleski (VK) | n/a | none |
| N. Karev (NK) | > 100 | none |
| D. Gruev (DG) | > 100 | none |
| Bratstvo-Edinstvo (BE) | < 30 | none |
| Lirija (L) | < 30 | none |
| B. Sabani (BS) | < 50 | none |
| Senic (S) | 30 | none |
| J. Sandanski (JS) | n/a | partial |
| Suto Orizari (SO) | n/a | none |

*Source:* Based on data in Vlcevski (1984, p.66)

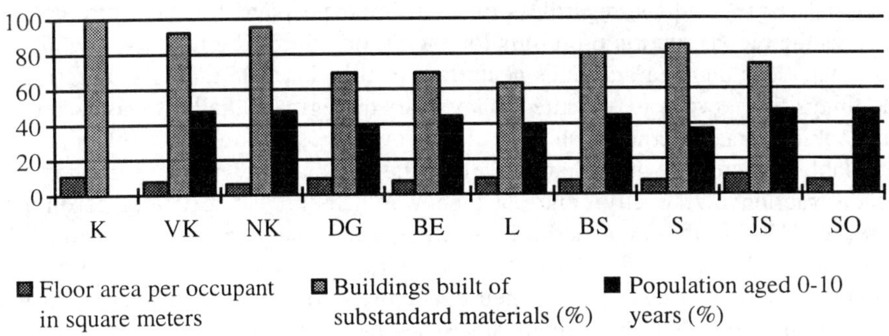


**Figure 6.16 Settlements with lowest standards in Skopje in 1981**
(note: all data were not available for Keramidnica and Suto Orizari)
*Source:* Based on data in Vlcevski (1984, p.66)

in the republic has been concentrated in Skopje. Therefore one of the city's priorities is to resolve this problem, if it is to achieve a safer and less vulnerable environment.

It seems that the uncontrolled migration to Skopje contributed significantly to difficulties in housing provision for the migrants from the rural areas, and led to an increase in Wild Development. Furthermore, this led to the transformation of the rural latent unemployment into actual urban unemployment, which is more difficult to deal with. The root of the problem rests in the immediate post earthquake period when people were encouraged to repair and rebuild without waiting for the required permits, as the officials were afraid for the future of their city and wanted citizens to return as soon as possible. These structures still exist and they have even multiplied. About two- to three-hundred illegally built dwellings or extensions are torn down annually, but the housing built or repaired immediately after the earthquake has been allowed to remain (Greene, 1987).

*Financial problems*

Today, 33 years after the earthquake, the planned city centre, especially the City Wall and the City Gate developments, has been only partially realised. The Transportation Centre, which was to be part of the City Gate, stands isolated from the city centre and the City Wall, because there have not been sufficient resources to complete the development of the land between.

Limited resources were also a problem in repair and reconstruction of older buildings, particularly those in the Old Market area and the residential areas on the left bank of the Vardar river, that were damaged during the earthquake and inadequately repaired afterwards. Both the Old Market area and the adjacent residential areas consist mainly of small, private shops and houses. Many repaired their homes and shops themselves, sometimes using substandard building materials or techniques, regardless of redevelopment plans for the entire area as well as the reconstruction proposals for each shop. Some did not have the money to repair their damage buildings according to the plan, while others preferred to do things their own way or waited many years to begin. Initially it was extremely difficult to obtain adequate funds for the redevelopment of the left bank, since the available resources immediately after the earthquake were put into the reconstruction of the city centre, which was heavily damaged. According to Greene (1987), the city authorities were particularly concerned about the materials and techniques used in the reconstruction from a fire and earthquake safety standpoint. Regardless of their concerns, Jordanovski (1993) believes that officials are partially to be blamed, since they have allocated most of the financial resources to the reconstruction of the right bank of the Vardar. In his opinion, it might have been better if the financial resources spent on the new Aerodrom housing complex development for 100,000 inhabitants, had been diverted to reconstruction of the existing areas on the left bank of Vardar.

*Transport and pollution*

The post earthquake developments led Skopje to become a linear city which had doubled in length from 12 to 24 km and in width from 6 to 12 kilometres respectively, (Jordanovski, 1993). The city sprawl and low density growth resulted in an overcrowded, inefficient and expensive transport system, entailing long journeys and environmental pollution (Greene, 1987; Jordanovski, 1993). The city growth surrounded several major factories (steel plant, cement factory etc.) that were originally located outside the city boundaries, thus contributing to further environmental deterioration. The transportation chaos was exacerbated by the insufficient parking provisions and underestimated growth of private car ownership, made even worse by the fact that in most of the new housing developments the planned underground parking and garages were never built. In addition, the existing railway line from the city centre to the Gorce Petrov suburb was removed in anticipation that the public bus service with later introduction of trams would meet the needs. The attempts of the local authorities to introduce trams in late 1980s failed, as on the local referendum the citizens of Skopje rejected the proposal for additional local income tax to finance the project.

*Cultural context*

According to Coburn and Spence (1992), disaster assistance policies that focus on assisting the reconstruction process by supporting rebuilding activities initiated by the community and allowing the community to control the reconstruction process, have proved to be more successful than outside agencies providing shelter. In Skopje, the international and national technical experts visualised major opportunities in rebuilding and re-rationalisation of the city, neighbourhood layouts and building improved housing with modern facilities, while the ambitions of the affected communities were more modest: to rebuild their familiar neighbourhoods and to re-establish their houses, streets and lifestyles. Their visualisation of the reconstruction was based mainly on what was there before and not what was going to be. But it was easier to plan and execute a large-scale reconstruction project centrally than to work in consultation with large number of communities.

The result of the international co-operation with all its benefits created problems of cultural 'appropriateness' in Skopje. The new structures and the rapid construction completely changed the outlook of the city. The new development broke down the compact urban tissue leading to a loss of many significant cultural and other city's identities (Arsovski, 1988). According to Korobar (1995), the 1963 Buchanan's Report ignored the already existing hierarchical traffic network and emphasised the City Wall road traffic, causing disruption of the newly introduced pedestrian routes. Furthermore, the earthquake was taken as an excuse to start building a new collective memory about the city, based on imitating city walls with banal modern administrative and residential buildings or by replacing city gate images with high rise buildings. These trade marks of the 'symbolism' have led to a lack of sense of history and continuity and to collective 'amnesia' about the city among its citizens (Golubovski, 1995). The city with its redefined spaces and with the introduction of an imported concept, ceased to represent a product of evolution and continuity. The rebuilding of Skopje raises therefore, the fundamental question of the appropriateness of a 'modern architecture' as opposed to an architecture evolving within a local tradition to satisfy local needs (Davis, 1975).

*Reconstruction attitude to architectural heritage*

Until the end of the 19th century, buildings were built using local materials and technology, and architecture was not usually devastated by unsuitable strengthening techniques in post earthquake reconstruction. The past experience proved that the traditional antiseismic methods, that could be traced back to biblical times, were reasonably effective - if they were used. If they were not used it was mainly for reasons of economy. Traditionally, local materials and regional construction techniques corresponding to the types and styles of architecture were used to reduce earthquake damage and to strengthen buildings. Fister (1988) argued that new building technologies changed architecture and were sometimes the only reason why some of damaged buildings had to be destroyed and not repaired. The latter was exactly what happened in Skopje after the earthquake. No attempts were made to use the traditional methods to restrengthen damaged buildings.

After the 1963 Skopje earthquake, all the antiseismic legislation and regulations recommended only methods developed for new buildings, based mainly on reinforced concrete technology. The new techniques did not consider the values of originality of materials and constructions to be important, but they were developed only to increase the earthquake resistance of buildings. The care to protect buildings, including architectural heritage and monuments, from future earthquakes was such that the conservation experts concerns were ignored. The modern construction technology used materials that improved the building stability but at the same time damaged the original materials (e.g., reinforced concrete against lime, etc.). As the legislation did not suggest special methodology for the architectural heritage, the buildings had to be changed and not the method. This attitude damaged the integrity, identity and the originality of the architectural heritage and monuments that were strengthened by insertion of reinforced concrete beams and columns.

Similar things happened in the case of 1976 Italian-Yugoslav border earthquake, intensity IX-X (MCS), where whole villages and a number of vernacular buildings were lost because the use of modern antiseismic strengthening technology was insisted upon while the traditional, simpler and cheaper methods where ignored. As a result, in former Yugoslavia, some of the most beautiful villages (e.g., Breginj) were destroyed, whereas across the border, in Italy, the more traditional earthquake protection techniques were used without problems (Fister, 1988). Three years later, after the 1979 Montenegro earthquake, intensity IX (MCS), reinforced concrete was used again for strengthening of traditional stone buildings and special methods were used only on few monuments. According to Fister (1988), that led to extermination of the architectural heritage in stone.

The authorities were not able to foresee that external funds for Skopje will not be available forever. As a result, they did not allow the more affordable traditional restrenghtening methods to be used in the reconstruction of the older building stock. This, coupled with the postponement of the city centre reconstruction, led to serious consequences that are visible today.

*Delayed reconstruction*

People in some of the city centre areas designated for redevelopment were led to believe that their neighbourhoods will be reconstructed very soon. Residents were not permitted to conduct any improvement or extension of their property, beyond the basic immediate post earthquake repair. Most of these areas were built in traditional or neo-classical style (traditional timber frame structure or brick masonry respectively). As the reconstruction was rescheduled year by year, these once prominent and desirable neighbourhoods of Skopje became substandard areas in the city centre. Those who were able to afford new housing moved out, creating two-fold effects. Firstly, they created a pressure on the limited housing market, and secondly, they sold their properties quite often to migrants. The latter were prepared to pay more than the city authorities, just to become residents of the city and to be able to seek employment. Some of them came with their extended families, transforming these areas into neglected and overcrowded slums.

Even if the city wants to redevelop these areas now, it will have to provide alternative accommodation for a substantially larger number of people, because of the increased population in these areas. If the authorities opted at the beginning for proper repair with structural strengthening of these buildings, with use of traditional techniques, they might have been able to create an earthquake-safer environment more economically, provide decent living standards for the inhabitants and give them the option to redevelop those areas when financial possibilities arise. One of the areas that has continued to deteriorate since 1963 is Madzir Maalo, the first example of orthogonal urban planning in Skopje from the Ottoman period, planned by Isa-Beg in 1878 (Arsovski and Taskovska-Arsova, 1988). The district is designated for demolition and development of the Tange's City Gate. The beginning of the reconstruction of the area was planned for 1985 when tenders were invited for the Jugobanka Headquarters that was never built, as the federal government banned all new non-productive development to ease the deepening recession. Now the 'green movement' advocates preservation of what is left of the old area, which in most other European countries would have been designated a conservation area years ago. Unfortunately the neglect of the area in past years raises doubts about its preservation without substantial financial resources, which are not available at present.

**Reconstruction after 1991**

It is difficult to assess the reconstruction achievements of late. Nevertheless, the construction activities that are going to take place in Skopje in the 1990s will be very important for the level of earthquake safety at the beginning of the next century. So far, the Republic of Macedonia has escaped the conflict in former Yugoslavia, but the war and the danger of the conflict spreading south into Kosovo and the Republic of Macedonia, has crippled the country's economy and reduced the Gross National Product per capita by half, to $771 (Norton, 1994). Despite government efforts to improve the economy, the country has been heavily affected by the UN sanctions on trade with Serbia and Montenegro, by the border blockade by Greece and by the reduction of trade with the rest of Eastern Europe and the former Soviet Union. The future of the country now rests with the international financing institutions, such as the World Bank and the International Monetary Fund, and their support is essential for successful reorientation of economic activity. Investments are needed in order for the Republic of Macedonia to maintain its economic and political stability, as well as to continue the post earthquake reconstruction of Skopje.

Major political, economic and social transformations have occurred in a matter of months. At the moment, because of the speed of changes, the government is not in a position to adjust and organise its responses accordingly, thus creating a form of legal vacuum that may undermine all the measures undertaken so far to increase Skopje's earthquake protection. It is an imperative that earthquake protection be addressed as soon as possible, with full awareness for the consequences of any delay.

Today it is very important that the construction standards and the enforcement of seismic and other building codes are not affected by the present situation, since periods of political and economic instability are characterised by the deterioration in implementation of development control and building standards, leading to potentially higher risk areas and buildings. Although it is difficult to assess the current state, it seems that the planning and construction standards in general have deteriorated in Skopje since 1991.

## Learning from Skopje

After 33 years, the residential post earthquake construction proved to be the largest reconstruction effort in Skopje. The Tange's City Gate and Wall have been scaled down and only partially implemented, while the Transportation Centre incorporated only few of the intended functions. The final reconstruction stages of the Plan have yet to be finished (Lagorio, 1990).

The reconstruction is a difficult process due to the resources and time required and, because of the foresight needed in many of the early decisions. Authorities should have the ability to make long term decisions at the expense of short term political and economical gains. Skopje is, partially, a remarkable effort because of this ability (Greene, 1987). Development was postponed in the city centre to allow time for architectural competition and was guided away from certain areas because of the potential seismic risk. The relief operation in Skopje is probably one of the most efficient on record, as minimum aid was wasted, and the authorities, (helped by the presence of the international emergency organisations), had very clear priorities. They avoided the waste of temporary housing but there are serious doubts whether the very rapid initial planning decision on the location of new settlements was correct (Davis, 1975). The decision to create new settlements with permanent housing immediately after Skopje's earthquake, coupled with the postponement of the reconstruction of the city centre, is partially to blame for the state of reconstruction achievement today. The reconstruction funds were dispersed and the over investment in heavy industry was a financial misjudgement, since this investment did not generate the expected income to finance the remaining reconstruction; instead generating more pollution and financial problems. Nevertheless, the reconstruction of Skopje still represents a successful approach in many ways, and is a remarkable example of efficiency and international co-operation following a disaster. It is especially important for the implementation of basic seismic safety features in its original concept (Lagorio, 1990).

The seismic considerations in the post earthquake reconstruction of cities date from 1668, when the Dubrovnik Senate's Act for Reconstruction of the City following the 'Big Shaking' of 1667 (Krstic, 1987), which led to the rebuilding of the Adriatic seaport that has survived many earthquakes since. Planning and architectural antiseismic measures were also adopted in Noto, Italy, during the eighteenth century reconstruction that followed the 1693 Sicily earthquake. After the 1755 Lisbon earthquake, the Marquis of Plombal introduced a reconstruction plan that incorporated some urban planning elements (regular block dimensions for easier access-egress, recommended street width and distance between buildings, limited height of buildings, etc.), to make the city safer during future seismic events. Recently, after the 1976 Tangshan earthquake in China, and the 1988 Armenia earthquake, urban planning strategies were reintroduced in the reconstruction process to reduce future earthquake hazards (Coburn and Spence, 1992; Lagorio, 1990).

Cities remain vulnerable to earthquakes because much of their building stock has been constructed before earthquake engineering was developed or not fully upgraded to take account of the latest knowledge (Lagorio, 1990). Nearly 75 per cent of the existing dwelling stock's floor area was lost in the 1963 Skopje earthquake and in the post earthquake period more than eighty-thousand new dwellings have been built. In 1991, only 20 per cent of the dwellings complied with the latest antiseismic code, 41 per cent with the outdated codes, while the remaining 39 per cent did not incorporate recognised earthquake protection measures. Thus, a large-scale approach to the problem is necessary, especially since future growth and change are expected. This includes consideration of open spaces, land-use planning, circulation systems, access/egress routes, building typology, and implementation of a general plan. The seismic safety element was introduced in the 1965 Plan in recognition of the earthquake hazards existing in Skopje and an implementation of comprehensive urban planning procedures was used to influence the reconstruction process and growth, and thereby reduce the vulnerability to future earthquakes.

The results of the analysis indicate that in the 33 year period of post earthquake reconstruction in Skopje, the recommendations in earthquake hazards mitigation have not been fully implemented, thus creating a potentially more earthquake vulnerable environment, regardless of the initial intention of hazard mitigation through reconstruction. There is a need to reduce future vulnerability and increase seismic safety in Skopje, since it is located in an active seismic region with possible earthquakes of an expected intensity up to IX (MCS). According to the 1965 Plan, densities in Skopje were reduced by enlarging the city's urban zone three times to accommodate a similar number of people and activities as before the 1963 disaster. Since the 1980s, these 'reduced densities' are being adjusted, by adding more storeys and building smaller flats, to accommodate the rapid, post earthquake growth. Thus, it is more difficult to comply with the original intent of low densities to minimise seismic vulnerability (Greene, 1987).

Several recent events raise serious concerns about the potential earthquake vulnerability of Skopje and the republic. The 1990 Gevgelija earthquake (magnitude 5.6, intensity VIII (MSK), with its' epicentre in northern Greece 25km away from Gevgelija) caused damage to 2,344 buildings (201 of them public buildings, including schools, medical facilities etc.) in Gevgelija and Valandovo (Milutinovic *et al.*, 1995). The greater level of damage was observed in buildings damaged earlier, in the 1978 Thessaloniki and 1985 Negotino earthquakes, and that had been inadequately repaired. The direct losses of the earthquake were estimated at $12 million. The 1994 Bitola earthquake, magnitude 5.4, intensity VII$^{+}$ (MSK), caused damage to 12,175 buildings in Bitola, Demir Hisar, Ohrid and Resen, and generated direct losses of $54 million or 3.4 per cent of the GNP of the Republic of Macedonia for 1993 (Milutinovic *et al.*, 1995).

After the privatisation in 1993, the two largest insurance companies in the republic, Makedonija and Vardar, excluded the earthquake risk from all their insurance policies. As a result, at present, no earthquake risk is insurable in the republic. This could easily create serious problems during future earthquakes. Even before 1993, when the earthquake risk was insurable, it used to cover only the structural but not the non-structural damages. With $711 GNP per capita, $1,300 million government debt towards the hard currency depositors, $1,200 million external debt (1994), real Gross Social Product growth of -7.2 per cent (1994), and 30 per cent unemployment (1996), the 2.1 million population of the Republic of Macedonia cannot cope even with smaller earthquakes like the 1994 Bitola earthquake, not a disaster such as the 1963 Skopje earthquake with estimated direct losses, at 1993 level, of $3,000 million (Dorgan, 1995; Financial Times, 1995; Jordanovski, 1993; Norton, 1994; Vesilind, 1996).

The key local players in the post 1963 Skopje earthquake reconstruction process can hardly be identified now. The City Planning Department is almost non-existent, while the Institute of Town Planning and Architecture is fighting for its survival, (patiently awaiting to move from their prefabricated office building into permanent premises since 1963). In 1975 the Faculty of Architecture removed from the curriculum the lectures in Geology, that used to incorporate the basics of seismology. The only institution that has maintained its expertise and reputation throughout the years is the Institute of Earthquake Engineering and Engineering Seismology. Now unfortunately, the institute is a target of a cost cutting exercise by the government, aiming at support to only a fraction of the present faculty. The republic and local government seem more preöccupied with foreign affairs and domestic inter party politics than with handling the issues of earthquake protection. At present few events can play a crucial role in the earthquake protection for the next millennium: the announced creation of the Republic Institute of Spatial Planning, the preparation of the Republic Spatial Plan and the preparation of the Republic Development Plan. If they appropriately incorporate and address the issues of earthquake risk and protection, with sensible implementation policies, then there is a chance of maintaining an earthquake-safer environment as implementation of earthquake protection measures, through planning and design, is more rational and efficient than the assessment and mitigation of earthquake consequences.

**Acknowledgements**

I am grateful to Prof V.-B. Music, Mr L. Erdel-Jan, Mr A. Pomonis, Mr N. Ray, Prof B. Kolev, Prof Z. Milutinovic and my wife, Dr E. Mukaetova-Ladinska, for their support throughout various stages of the research. This chapter is based on a research funded by the Republic of Macedonia Science Research Council, the Anglia Polytechnic University and my parents - architect Vasilka Petrovska-Ladinska and the late Prof Boro Ladinski to whom I dedicate this work.

# 7 Housing and resettlement in post-war Liberia

*Adenrele Awotona and Mulbah G. Johnson*

## Introduction

Armed conflicts, involving two or more countries, as well as domestic wars have, since World War II, brought about untold human suffering and massive destruction to millions of people world-wide. Although accurate figures are difficult to estimate, Walker (1985, p.68) points out that the International Federation of Red Cross' and Red Crescent Society's assistance, costing some 23 billion Swiss francs (EUC 270 million) annually, is currently going to some 19.4 million people world-wide in international relief operations; 6.6 million of these are refugees or displaced people. This figure excludes the amount being spent annually by the United Nations on peace keeping and relief efforts. What has become alarming is not so much the number of victims and the amount of financial resources involved in disaster relief, but the continuous failure of relief operations to achieve complete recovery.

While in many cases the provision of housing after a prolonged war may be inevitable, such provision often neglects the fact that in the event of disaster people lose 'homes' - not mere buildings. These are the reflection of their cultural beliefs, social and economic practices and aspirations as well as their relation to the community and the environment. Recovery from a disaster can be a complex process, especially in the case of a war situation where, along with the destruction of the physical structures, there is also a breakdown of the political and social institutions that are essential for reconstruction. Whatever the nature of the disaster, the chances of complete recovery are largely dependent upon the extent of the damage created as the result of the disaster, and the ability of the reconstruction programme to respond adequately to the real needs of the disaster victims.

Researchers and disaster relief workers alike, have been baffled by the recurring failure of most post-disaster relief initiatives and subsequent reconstruction programmes. Cuny (1983) when discussing the persistent failure of relief and reconstruction programmes, argues that post-disaster reconstruction programmes, especially those in developing countries, have, in most cases failed because they were modelled on the predominantly centralised planning system adopted in Europe after the Second World War. As such they lack the necessary financial resources and technical knowledge for implementation. It has become a standard practice to supply certain goods to every disaster area irrespective of disaster type or nature. Among the best known items are tents, clothes, high-protein foods, milk powder and blankets. Each of these items of course, may be useful in a specific situation. However, it appears that aid organisations are putting greater emphasis on having these items ready for shipment to disaster areas, rather than developing appropriate mechanisms for adequate post-disaster needs assessment.

In the absence of a national plan to guide relief and rehabilitation programmes, the international relief operations in Liberia are costing a lot and are of little benefit to the victims. There is over-lapping of functions amongst relief agencies. The activities of most aid organisations operating in the country are designed on the guidelines of their foreign donor organisations. They are neither accountable to the victims that they claim to serve, nor to the Transitional Government in Monrovia. Against this background there are strong fears among many Liberians that if this trend continues, future reconstruction programmes might also be designed along similar guidelines.

The aim of this study is to explore the theoretical issues surrounding post-war reconstruction and how these issues have manifested themselves in past reconstruction programmes in general, and housing and resettlement programmes in particular. This aim will be achieved by identifying the common problems which recur in post-disaster housing and resettlement programmes, investigating the extent of the housing problem in Liberia prior to the war; and examining the level of physical destruction and socio-economic disruptions being created as the result of the war. The analysis of the relationship between these objectives will provide a better picture of the issues of post-war reconstruction, and will thereby serve as a basis for making recommendations for the planning and management of housing and resettlement programmes in post-war Liberia.

## War as a disaster

In the words of Turner (1978)

> Mankind lives with the knowledge that its plans and intentions may be thwarted by disaster. Even when it thinks itself safest, it can never wholly discount the possibility that some unforeseen and destructive event may subject men to danger, or disrupt their orderly everyday affairs. Such events have

commonly been regarded as resulting from external and unfathomable force which could not be directly controlled, but only accepted.

The number of disasters have grown in recent times and they are affecting more people than ever before. Disaster now hits between 250 and 300 million people a year, excluding those caught up in war. Although annual figures vary widely, Walker (1995, p.68) explains that the general trend is upward at a rate of around 10 million a year.

Although armed conflicts and natural disasters display similar characteristics in many ways, including loss of lives and property, there are a number of differences regarding long term physical, psychological and economic impact. While Davis (1987a, p.8) generally regards natural disasters as a coincidence between natural hazards and vulnerability, disasters brought about as the result of civil conflicts are the result of deliberate imposition of hazard (man-made) on vulnerable people and environment. Having said that, unlike natural disasters, armed conflict can be anticipated and sometimes steps taken to minimise casualties through evacuation.

Important differences between the two types of disaster exists, and these affect reconstruction efforts. The first, according to Kreimer (1988), involves the time dimension and its effect on a country's institutions. Because natural disasters have a sudden impact, the institutions may be weakened by a large number of demands but still be relatively intact. Thus reconstruction efforts can be undertaken using existing institutions and helping them function more effectively. However, after prolonged wars, there is greater disruption to the entire society, and institutions are apt to be functioning poorly. Another problem, unique to the aftermath of war, is that over the years in which the fighting continues, people migrate or are forced to leave their cities; this displacement affects the availability of skills. In the same vein, Kreimer (1988) explains that the construction sector, which is critical to reconstruction once war ends, may be badly weakened, since it tends not to be very active during the war.

The destruction of war is not limited to only physical damages. The disruption caused by war can be viewed from different standpoints - political, social and economic. Besides physical destruction, which is measurable, social and economic disruptions which include separation from family members, separation from one's normal way of life, loss of business, and the loss of employment opportunities, are just as serious, if not more than the physical destruction. The extent of damage caused by war depends upon several factors: (1) the kind of war, (2) the kind of weapons used, (3) how long the war lasts, (4) the level of vulnerability of the civilian population, and (5) the social, economic and political condition of the affected area.

## The effects of the Liberian civil war on human settlement

The civil conflict in Liberia has been an enormous tragedy in the history of the Country. Every Liberian, regardless of economic and social status or tribal affiliation, has been affected by the violence, destruction and human rights abuses. The infrastructure of the country, its economy, social systems, moral and cultural values, are being severely damaged (Figures 7.1, 7.2 and 7.3). The infliction of atrocities by one group of Liberians on another is resulting in one of the most violent events in post-World War II history. According to Kesseley (1995, p.8), a US State Department Report on Liberia revealed that since 1989, when Liberia's population was recorded at 2.4 million, an estimated three-hundred-thousand people (12.5 per cent), most of them civilians, have been killed or wounded as a result of the fighting.

**Figure 7.1 A war damaged building on the main campus of the University of Liberia**

**Figure 7.2 A damaged electric power transformer**

**Figure 7.3 The unfinished Ministry of Defence building. One of the many Government projects abandoned since 1989**

*Population displacement*

Hundreds of thousands of Liberians have been forced to leave their homes at one point or another during the last five years. Out of the 1990 estimated population of 2.4 million people, The United Nations High Commission for Refugees (UNHCR, 1995), estimates that around 755,000 Liberians have been externally displaced as refugees in Sierra Leone, Guinea, Ghana, Nigeria and Cote D'Ivore. Another 1.1 million are displaced internally (Figures 7.4 and 7.5a & b). Others who possessed the necessary resources, fled to the United States or Europe in search of safety. Due to the relative calm prevailing in the Monrovia area, there is a constant influx of people in the city, increasing the population to 800,000 inhabitants, twice its pre-war level of about 400,000 (SBA, 1996).

**Figure 7.4 The Ministry of Foreign Affairs building. Home of some 3,000 displaced persons**

**Figures 7.5a & b The unfinished Ministry of Health building. Also home of some 7,000 Sierra Leonean refugees and Liberian displaced persons**

While displacement is traumatic and terrible for everyone, the impact on women is particularly harsh. Displaced women in Liberia are targeted for special abuses because of both ethnicity and gender. Rape is used as a form of torture and many women are suffering psychological and physical trauma as a result. This physical abuse faced by displaced people is just a fraction of the problems they have encountered. The living conditions in the camps are appalling (Figures 7.6a & b). There is a high risk of endemic diseases and malnutrition in many of the displaced and refugees camps. As shortages become acute, people have to do without basic items like food, medicine and clothing. The inadequacy of assistance or difficulties in distribution and delivery means that women have to sometimes provide sexual favours in order to feed their families. Some women are forced into prostitution as their only means of earning a livelihood.

*Housing demand*

The housing situation has become more problematic due to the hostilities and the destructive effects of the war. There are two major factors which have intensified the housing crisis in the country. Firstly, there has been a huge destruction of the housing stock in rural areas. Every town and village throughout the country was hit by the wave of destruction, leaving only a few of the pre-war rural housing stock in tact. This situation has increased the rate of rural to urban migration, thereby increasing the demand for housing in urban areas. Furthermore, the fact that the urban areas also experienced massive destruction of residential houses, has increased this demand further. According to Subah (1994), there were an estimated fifty-thousand buildings partially damaged or completely destroyed in Monrovia and its environs between 2 July and 29 November 1990. The renewed fighting between 15 October 1992 and 23 December 1992, resulted in an additional wave of destruction of houses and other properties, in and around the city. All of these factors as well as insecurity and inflation, have increased the price of and demand for housing in relatively safer parts of the country, and pushed up the cost of renting a housing unit, to the extent that it has become practically impossible for middle and low income groups to find suitable dwellings.

**Figures 7.6 a & b Voice of America (VOA) Refugees and Displaced People's Camp on the outskirts of Monrovia, home of some 26,000 persons**

**Table 7.1**
**Type of shelter occupied by displaced families in Voice of America (VOA) Displaced Camp outside Monrovia (sample size of 600 families)**

| Family size | Tent | Bamboo mat | Concrete | Zinc sheets | Mud and poles | Total |
|---|---|---|---|---|---|---|
| 1 | 1 | 2 | - | - | 2 | 5 |
| 2 | 1 | 5 | 3 | - | 3 | 12 |
| 3 | 13 | 4 | 16 | - | 17 | 50 |
| 4 | 17 | 21 | 26 | 3 | 72 | 139 |
| 5 | 22 | 31 | 24 | 2 | 59 | 138 |
| 6 | 10 | 24 | 16 | 1 | 39 | 90 |
| 7 | 14 | 10 | 14 | - | 31 | 69 |
| 8 | 7 | 10 | 9 | - | 15 | 41 |
| 9 | 3 | 9 | 3 | - | 10 | 25 |
| 10 | 1 | 2 | 5 | - | 9 | 17 |
| 11 | 1 | 2 | 1 | - | 9 | 13 |
| 12 | - | 1 | - | - | - | 1 |
| Total | 90 | 121 | 117 | 6 | 266 | 600 |
| Ave. (%) | 15 | 20.2 | 19.5 | 1 | 44.3 | 100 |

*Source:* Adopted from SBA (1996) *Monrovia Baseline Survey,* An E.U. sponsored survey conducted by Subah Belleh Associates,. Monrovia, unpublished

Attempts made by relief agencies to organise camps for the displaced, especially in and around urban areas, (using locally available building materials to build makeshift shelters), have made a contribution, but have not totally solved the shelter problems facing displaced people. SBA (1996) estimates that 30 per cent of households in and around Monrovia are occupying inadequate and unsafe dwellings. Although similar data are not available for the other major urban areas, preliminary assessments have shown that the post-war housing problems in Gbarnga, Buchanan, Greenville and Harper are similar to those being experienced in Monrovia.

**Table 7.2**
**Distribution of sleeping rooms by displaced family size in Voice of America (VOA) Displaced Camp outside Monrovia (sample size of 600 families)**

| Family size | 1 Bedroom | 2 Bedrooms | 3 Bedrooms | 4 Bedrooms | Above 4 bedrooms | Total |
|---|---|---|---|---|---|---|
| 1 | 5 | - | - | - | - | 5 |
| 2 | 11 | 1 | - | - | - | 12 |
| 3 | 41 | 4 | - | 1 | 4 | 50 |
| 4 | 111 | 16 | 2 | 2 | 8 | 139 |
| 5 | 103 | 25 | 3 | 2 | 5 | 138 |
| 6 | 69 | 13 | - | 1 | 7 | 90 |
| 7 | 42 | 19 | 1 | 1 | 6 | 69 |
| 8 | 29 | 6 | 2 | 2 | 2 | 41 |
| 9 | 20 | 4 | - | - | 1 | 25 |
| 10 | 10 | 7 | - | - | - | 17 |
| 11 | 7 | 3 | - | 3 | - | 13 |
| 12 | - | 1 | - | - | - | 1 |
| Total | 448 | 99 | 8 | 12 | 33 | 600 |
| Ave. (%) | 74.7 | 16.5 | 1.3 | 2 | 5.5 | 100 |

*Source*: Adopted from SBA (1996), *Monrovia Baseline Survey*, An E.U. sponsored survey conducted by Subah Belleh Associates, Monrovia, unpublished

With the lack of organised emergency housing provision, especially in Monrovia and Buchanan (the two cities with large displaced populations), and under pressure of need for shelter, displaced people in these cities have adopted one of two tactics: squatting on vacant land and building makeshift dwelling progressively, or occupying vacant buildings which already exist. Although no exact information is available about the number of people occupying vacated buildings throughout the country, a UNHCR registered displaced camp in the unfinished Ministry of Health building in Monrovia has approximately seven-thousand people. The vacated Foreign Ministry building is home to around a further three-thousand displaced people. Considering that there are around 10 major buildings being occupied in the Monrovia area, the number of displaced people living in vacant buildings is thought to be between 20,000 to 50,000.

*Social and economic disruptions*

The crisis has brought education to a complete state of paralysis and dismantled its structure and substance. The pre-war school enrolment of approximately 400,000 pupils, dropped to about 180,000 in 1992 (NARDA, 1992). School facilities have either been looted, destroyed or are being used to accommodate displaced people or to host military units. There is also an acute shortage of educational materials and furniture. Basic facilities such as workshops and laboratories are deteriorating. The crisis has also caused the prolonged separation of thousands of school age children from their families and consequently an increased incidence of deviant behaviours among the youth. More deviant behaviours are now exhibited, such as the disregard for family, community leaders, and school authorities; the lack of a sense of national identity or patriotism, large intake of alcohol, vandalism, obsession with guns etc..

The economy, previously based on iron ore, rubber, diamond and gold export remains stagnant. In the absence of export activities, internal commercial activities, including banking in Monrovia, are operating at less than half their pre-war capacities. Unemployment across all sectors is as high as 80 per cent. Massive displacement of civilians, wanton destruction and looting have all devastated the productive capacity of Liberia, despite its rich natural endowment and potential self sufficiency in agriculture.

Although official statistics are almost non-existent, Cole (1995, p.8) estimates that between 1990 and 1992, the cost of living in the country sky-rocketed to over 2,000 per cent of the 1989 figures, and is in 1995 dangling between 4,500 and 5,000 per cent. The Liberian dollar, which has an official value of Liberian$1 to US$1, is being traded on the black market at the ratio of US$1 to Liberian$55. Essential commodities have in the last five years, disappeared from the market, and the prices of those that have remained are multiplying by the hour. Major businesses, including banking institutions and mining companies, have been forced by the prevailing economic situation to either close down their operations or to operate at reduced levels.

*Infrastructure*

At present it is difficult to make a detailed assessment of the physical damage to the infrastructure of the country. Firstly, the fact that the entire country, with the exception of Monrovia and a few coastal towns, is controlled by one warring faction or another makes free movement outside of Monrovia almost impossible. Secondly, areas that were initially affected by the fighting during the early days of the war (1989-1991), and later experienced some form of physical reconstruction, have in recent times been bombed and destroyed. Gbarnge in central Liberia and Kakata are cities that fall in to this category. Thirdly, the decade preceding the war saw the decline in maintenance of public utilities. Essential national development projects were abandoned and allowed to deteriorate. This situation makes it difficult to distinguish damage caused by neglect from that being inflicted by the direct action of the war.

However, preliminary assessments carried out of infrastructure and public services in accessible areas have shown severe damage to public facilities. The hydroelectric power plant which supplied 70 per cent on Monrovia's power requirement was blown up in the early days of the war. As the result other services, that relied heavily on electricity, were also shut down. At present Monrovia receives only 10 per cent of its pre-war electricity supply, 6 per cent of its pipe-borne water supply, and 50 per cent of its pre-war telecommunications services (Liberia, 1993). Roads are also in a terrible state due to lack of regular maintenance during the last five years. In addition to the damage caused by fighting, major components of machinery in dormant factories are also being subjected to looting and vandalism. Electrical and telecommunication cables, water pipes, and heavy earth-moving equipment are being removed and exported as scrap metal.

## A synopsis of Liberia's pre-war housing situation

Most difficulties encountered during reconstruction, are due to the failure of post-disaster damage assessment schemes to distinguish pre-disaster housing problems, and those brought about as the result of the disaster. This situation has become a common phenomenon among the recurring problems of post-war reconstruction programmes. Therefore a better way of understanding the degree to which human settlement in Liberia has been affected by the civil war, and a subsequent intervention via reconstruction, will require a full knowledge of the housing situation prior to the conflict.

### *Pre-war housing policy*

Before the 1960s the Government's attitude to housing was based on a *laissez-faire* policy. The production and supply of housing was entirely left to the private sector. There was neither a financial institution with specific duty of financing housing loans, nor an institution for housing research and development programmes. The first step by the Government to develop a national housing urban development policy came in 1960, when the National Housing Authority (NHA) and the National Housing and Savings Bank (NHSB) were created by an Act of the Liberian Legislature to act as the Government's custodians for housing. These two institutions were empowered by law to plan, initiate and implement housing development programmes.

The approach adopted by the NHA centred on the commitment of budgeted public funds to conventional mass housing programmes, often targeted at low and middle income groups with approximately 30 per cent direct subsidy from the Government. There were later attempts made at producing a more sustainable housing programme by encouraging users' participation and accountability, through self-help community upgrading and a site and services project.

For reasons associated with both institutional and policy failures, the NHA was unable to live up to its task. During its 30 years of existence (1960-1990), the achievements of the NHA were the construction of a mere 5 low and middle income housing estates (Figures 7.7 a & b), in and around Monrovia, totalling 1,668 units; 1 site and services project comprising 515 serviced plots, 100 of which contained building cores, and a few slum upgrading projects in the centre of Monrovia (Mason, 1993). Most importantly, the activities of the NHA were never extended to other parts of the country.

*Housing quality and supply*

The extremely poor quality of dwelling in Liberia pre-dates the ongoing civil conflict. Even though national data are hard to come by, the result of the 1974 housing census revealed that poor housing, rather than limited housing, was the main housing problem in Liberia (NARDA, 1992). For example, the 1974 census showed that 1.39 households inhabited a given house. However, while the effective housing demand in the late 1970s might have been the need for improving already existing units, rather than for additional ones, the rapid growth of the urban population might have changed the situation. For example, Monrovia's population grew from 204,200 in 1974 (Lacey and Owusu, 1988, p.44), to 400,000 in 1989 (Subah, 1994). This means that the need for additional housing also became part of the problem.

The census of 1974 also showed that the quality of the housing was very low. Eighty-eight per cent of the total housing stock was considered sub-standard; 20 per cent was considered hazardous and needed massive repair or demolition; 83 per cent did not have any utilities whatsoever; and only 6 per cent of all households had pipe-borne water, electricity and flush toilet. Forty-three per cent of all households also shared kitchen and toilet facilities (Liberia, 1984). According to NARDA (1992), research information, prior to the outbreak of the civil war, established a further worsening crisis in the housing sector for reasons which included the rising cost of building materials, the low level participation of the private sector in housing and urban development, the general non-participation of NGOs in housing, and the inability of the Government sponsored programmes to adequately respond to the deteriorating housing stock in Liberia. Indeed of the housing stock 11.7 per cent was provided by the public sector, 18.3 per cent by the private (formal) sector and 70 per cent by the informal sector. Data gathered by NARDA (1992) shows that only 1.9 per cent of the housing stock of Monrovia was provided or partially funded by the public sector.

**Figures 7.7a & b Matadi Housing Estate. One of the five public housing estates owned by the National Housing Authority (NHA)**

The quality of housing in informal settlements, particularly in Monrovia, varies from extremely deplorable to fairly acceptable (Figures 7.8 a & b). The availability of services such as water supply, electricity supply and sewerage systems have not kept pace with the rapid growth. Only 40 per cent of the population of Monrovia had access to pipe-borne water prior to the war in 1990. Many residents, especially in low-income communities, lacked these services or paid a large proportion of their small incomes for inadequate services. According to NARDA (1992), in several housing surveys conducted by the NHA between 1970 and 1978 in Monrovia, residents of low-income communities ranked public toilet, clinic, and street lighting as priority needs followed by water supply, garbage collection, street paving, school and improved transportation, all of which were either in deplorable conditions or inadequately supplied. The construction of new dwellings trailed much farther down the priority needs ladder.

*Constraints in the housing sector*

*Institutional* The lack of a coherent national housing policy for clearly defining the overall objectives and responsibilities of the various participants (Government, private sector, NGOs, community groups etc.) in the housing delivery system, and to provide guidance to the process of how housing should be delivered, resulted in the general state of confusion, with little or no organised and reliable information on housing and urban development. There were overlapping and duplication of responsibilities and functions amongst the various agencies and institutions. The Ministries of Internal Affairs, Public Works, Finance, Presidential Affairs, and Lands, Mines and Energy, as well as the National Housing Authority, General Services Agency, National Social Security and Welfare Corporation, and Monrovia City Corporation all shared bits and pieces of the responsibility for housing and urban development with little or no co-ordination whatsoever. This situation, coupled with inadequate community structures through which community efforts could be harnessed and channelled into more productive and sustainable housing development activities, rendered the national institutional framework for the provision of housing ineffective.

*Land tenure* There exists the problem of unreliability and ineffectiveness in the national land registration system. This ineffectiveness of the national land administration and management system was due principally to the failure of housing, and housing-related agencies, and the Government to recognise land as an important economic resource, and thus prioritise its management. Furthermore, the existing real estate tax laws encouraged land speculation and inadequate development, by penalising real estate developers with increased taxes and encouraging land speculators with little or no taxes. The reluctance of Liberians to accept communal ownership in urban areas, the unwillingness by underdeveloped urban land holders to sell to developers, and the lack of

**Figure 7.8 a & b   A low-income housing area in Monrovia**

condominium title laws, have all contributed towards retarding the rapid development of housing (NARDA, 1992, p.54).

*Construction and zoning* The lack of construction standards appropriate to the application of indigenous materials and low-cost technologies had a negative impact on the housing industry. Conventional and imported building materials and technologies made housing production very expensive. The existing standards and zoning laws, which according to NARDA (1992, p.55) have not been revised since 1956, are unrealistic and socio-culturally irrelevant for the conditions of Monrovia. These standards and zoning laws have been biased towards foreign designs and foreign materials which are often inappropriate to local conditions and hardly affordable.

*Housing finance* Permanent amongst the problems of housing delivery in pre-war Liberia is the issue of finance. While there has been no real access to credit for the largest supplier of housing, the informal sector, the formal housing sector also witnessed mass withdrawal of investment in housing by financial institutions. The National Housing and Savings Bank, which was specifically created to fund housing programmes, abandoned its role for commercial banking because of the difficulties it encountered in ensuring debt repayment.

This withdrawal of financial support from the housing market by the NHSB and other financial institutions was partially due to the absence of appropriate legal frameworks, that would enforce loan repayment, while at the same time protecting the mutual interest of financiers and beneficiaries. Another possible reason for the withdrawal of investment from housing is that financial institutions did not carefully study the nature of the housing situation in the design of their portfolios, and tended to favour larger and so-called prestigious projects which ended in financial disasters, thereby creating an adverse impression of the housing market.

The relatively poor performance of the housing sector was the result of a combination of factors, ranging from the adoption of a housing delivery system that was financially dependent on Central Government, the lack of involvement of the NGO community in housing development, and therefore a failure to develop housing as a viable economic and social enterprise; to the lack of a clearly defined set of objectives by which the role of housing in the formulation of strategies for national development could be established. Given the above, one can easily conclude that the situation of housing in pre-war Liberia was not optimal, and that damage done to the housing stock by the war should be considered as an addition to a pre-existing problem. The fact that an already bad situation has been made worse, necessitates the need for reconstruction plans and programmes aimed at addressing both the pre-war housing problems as described above, and those that are being created as the result of the civil war.

## Recommendations for developing a national housing and resettlement plan

It is inappropriate to conduct comprehensive reconstruction in a continuing war situation (Zargar, 1989), there can be no effective reconstruction before a final peace agreement, or at least not in the areas controlled by the warring factions. Therefore, it would be premature to recommend specific strategies for developing a national housing and resettlement plan, in the absence of total peace, bearing in mind that the present conditions in the country, as described earlier, could probably change between now and the end of the war. Having said that, it would also be inappropriate to subject all the planning for reconstruction to a final peace agreement. Barakat (1993, p.48) argues that housing and resettlement planning should not necessarily be linked to total reforms, as waiting for - or linking construction planning with, major economic or political reforms (such as legislative changes in land tenure, taxation or local democracy) mean losing the impetus for change which exists in the immediate post-disaster period. General policy-making guidelines for reconstruction planning can be worked out at this stage. These guidelines can then serve as a starting point for a preliminary strategic plan which will facilitate detailed planning of the subsequent physical rebuilding process, socio-economic restructuring, and political reforms.

### *Post-war housing and resettlement planning*

There is no single procedure or definite rule for housing and resettlement planning. However, a successful plan is one that places the welfare of people at the centre of the reconstruction process, rather than economic or technological issues (Davis, 1986). Improper planning and implementation of post-war housing and resettlement programmes in the past have led to failure. These failures have been either in the form of the inability of implementing organisations to complete projects, or that programmes initiated were of no real benefit to the disaster victims. Planning in the specific circumstances of reconstruction after war must take into account the actual needs of the victims, as well as the need to find prompt solutions to meet the immediate socio-economic needs of devastated communities. Post-war housing and resettlement plans should take into account issues that are not only geared towards the replacement of lost houses but also the creation of lost social environments.

The fact that reconstruction planning will be inhibited by pre-war deficiencies means that recognising and addressing some of these deficiencies, is crucial to the success of reconstruction. A strategic analysis of pre-war deficiencies must be made, and programmes aimed at tackling those problems or removing constraints developed. Priority must be given to those constraints that might impede the effectiveness of the housing and resettlement programme if left in place. However, steps should also be taken to avoid bulking down housing and resettlement processes by attempting to solve all the pre-war problems before initiating implementation. In terms of what should be the first priority in reconstruction, the provision of shelter and support for the sick, the homeless and the unemployed must be the top of the list. However to achieve this, utilities will need to be restored and that in turn, requires the restoration of the construction industry and the rebuilding of national and local economies.

It is likely that conventional planning procedures and building approval mechanisms will need urgent reappraisal to make them applicable to the challenges of reconstruction. Steps must also be taken to avoid drawing up rigid master plans for determining reconstruction activities. Rowland (1995, p. 32) argues that master plans take a long time to prepare - and can be out of date when completed, or at least of little value to agencies that have to take key decisions in the interim. Secondly, it is important to understand that even with the best possible plan of action worked out, it is likely that the reconstruction programme will change significantly from time to time. Recognising this in advance, and building-in suitable flexibility, can save a tremendous amount of time and resources, and allow the programme to react to new situations or opportunities as they arise (Cuny and Tanner, 1995).

*Institutional capacity building*

After prolonged wars, there is a disruption to the entire society, and institutions are apt to be functioning poorly. Many of the staff may have changed, and probably no training has occurred during the time of the conflict. Therefore, plans should be made for the rebuilding and restructuring of various national institutions if they are expected to effectively cope with the challenges of reconstruction. The immediate prospects of physical reconstruction will depend on the extent to which the construction industry has survived the war, and how quickly it can be revived. At best it will have been depleted of labour and materials. The supporting professionals are likely to have been greatly depleted. Rebuilding the construction industry and restoring its financial and material resources will have to be a priority.

## Conclusion

While many resettlement programmes have run their course without problems and have resulted in a total return of all refugees and displaced people and their subsequent effective reintegration into their home regions, in other cases, repatriation has proved to be most difficult and problematic (Rogge, 1994, p.14). Caution must therefore be practised when developing plans for repatriation and resettlement of displaced people and refugees. Repatriation and resettlement require co-operation from the affected community.

Generally, post-war housing programmes have fallen short of expectations as far as meeting the needs of the victims are concerned. This has been partly due to the conflicting perception of housing needs by both the victims and the intervenors.

Disaster victims on the one hand regard housing, not only as a means of physical protection from the harshness of climate and environment, but also a means of cultural, social and economic expression. Intervenors on the other hand, tend to treat housing as a quantifiable object. Therefore, quantitative surveys, widely used for needs assessment in relief and reconstruction programmes, have mostly been ineffective and inappropriate. It has also been pointed out that disaster relief programmes, while designed to help the victims, tend in fact to reflect the needs of the donors. This is a result of attempts by relief agencies to please their donors, as well as the lack of accountability of these agencies and donors to the victims.

The inadequate assessment of housing needs after disasters such as war can be attributed, amongst other factors, to the lack of familiarity of the assessors with the local situations, and the lack of understanding of appropriate techniques for damage and needs assessment. The fact that most disaster needs assessments are done by outsiders, the lack of knowledge of pre-existing housing conditions often makes it difficult, if not impossible, to distinguish between disaster-related needs and pre-existing housing shortages. Consequently shelter requirements may be overstated, attributing residual housing deficiencies to the disaster. Furthermore, Intervenors traditionally regard disaster victims as helpless, people who are unable to help improve their physical conditions. This view is the result of a myth that grief traumatises disaster victims to the point where they must be led into activities in order to save themselves. In addition to the above stated notion, the conventional methods used in data collection are usually inefficient in the chaotic conditions of immediate post-war periods. The use of these conventional methods, coupled with the lack of sufficient input from the victims, often results in the failure of the assessment to distinguish between housing 'wants' and 'needs'.

To avoid such occurrences in the situation in Liberia, popular participation in the Post-war Housing and Resettlement Programmes must be encouraged. Resettlement and reconstruction activities are more effective when they involve people and communities at the local level. It has been proven that local control improves public accountability and strengthens local capabilities. Where decisions are centralised and regional traditions ignored, results are often unsatisfactory (Johnson, 1995, p.67). Therefore, allocation of roles to a wide range of governmental departments, as well as to non-governmental organisations and local community groups, will ensure fast and effective implementation. Popular participation in resettlement planning is also vital, since it is unrealistic to expect people to fit in with predetermined plans.

At every possible point, the local population that have survived the war must be involved in the planning process and rebuilding activities. Experience has shown that there are dangers in responding to the challenges of reconstruction by bringing in external expertise to reconstruct peoples' homes and settlements. External help will almost certainly be needed, but it must be provided in a support role to strengthen what already exists rather than to replace or dominate indigenous efforts. Reconstruction should be a local enterprise, locally controlled (Green, 1995, p.38). However, this need for maximising local input must not be misconstrued to mean compulsory exploitation of local communities, but rather as a tool for utilising local resources to the best effect. The success of a repatriation and resettlement programme will also depend on an appropriate land tenure policy, and secure access to employment opportunities, services and infrastructure.

# 8 Typhoon resistant housing for the poorest of the poor in the Philippines

*Diane Diacon*

## Introduction

The Philippines Archipelago is one of the most disaster prone areas in the world, regularly affected by typhoons, landslides, earthquakes and volcanic eruptions.

One of the major causes of destruction in the area, both of life and property, are typhoons. Somewhere between 5 and 9 occur each year, causing widespread damage. Flimsy self-built shanties and huts are tossed in the air like straw by the wind. The concomitant flooding frequently renders towns and villages inaccessible for days or weeks on end, because of landslides and flash floods. The cost in human terms of typhoons, from 1970-1982, was the loss of 3,867 lives with a further 6.2 million people directly affected and damage to property estimated at $683 million.

Particularly susceptible to the typhoons are the traditionally built houses made from locally available materials such as nipa (leaves). This is the only form of housing available to the poorest families and has no resistance at all to typhoons. Families living in such dwellings expect to completely rebuild their home once, if not twice a year, and a family's entire resources are concentrated simply on keeping a roof above their head. As soon as one inadequately constructed roof blows away, another is built, with the result that families never have time to improve their standard of living by planting gardens or improving their houses, or by establishing small-scale agricultural or manufacturing enterprises. All efforts are concentrated upon the struggle against the capricious elements. Efforts to rebuild such houses are mainly on a self-help ad-hoc basis, supplemented by extremely limited assistance from the government and periodic aid from international aid organisations.

Typhoon Sisang in 1987 proved the turning point. Just like so many other typhoons it rendered many thousands of families homeless - 490,000 families were affected, 207,000 houses completely destroyed. Rather than use the international aid available to provide yet another temporary solution to the problem, the government decided to try to seek a more permanent solution. It sought to break the cycle and find another way of housing these desperately poor families whose homes were being repeatedly destroyed. It was decided to design a low-cost house which was resistant to typhoons and which the people could build themselves while still utilising local materials. Only in this way could the cycle of devastation, such a limiting factor in the development of these lives and communities, be broken.

**Starting the Core Shelter Project**

The Department of Social Welfare and Development (the DSWD) is the government department in the Philippines which has the task of providing emergency relief assistance and social services to victims of disasters. Its aims are threefold: to help people cope with the crisis, to meet their immediate basic needs and longterm, to help them to get back on their feet and on with their normal lives.

Financial resources as always are limited, and it was decided that the most cost effective way of spending the international disaster relief money for Typhoon Sisang was on the design of a low-cost house capable of withstanding typhoons. The DSWD invited the highly experienced and respected Asian Disaster Preparedness Centre (ADPC) to advise.

Dr Satyendra P. Gupta of the ADPC visited the affected area in late 1987 in order to inspect the damage, assess the causes of weakness in the present structures, provide advice on reconstruction and make recommendations to improve the situation in both the long and short term. The design brief required that the unit be capable of being built of locally available materials using the traditional building skills of the local people. Any typhoon resistant design features included had to be simple and acceptable to local traditions if the project was to be successful.

With the advice of the ADPC the DSWD developed a new design for a shelter. It was known as a core shelter, since it provided a basic typhoon-secure unit or core into which families could retreat in the event of a typhoon, together with any stock and other possessions that they wished to protect.

Once the basic design had been developed it was pilot tested in the three regions of the Philippines which are struck most regularly and heavily by typhoons. Four-hundred and fifty units were constructed during 1988, providing homes for some of the victims of the Sisang disaster. These units faced two typhoons later that year, with wind speeds reaching 160 and 175 km per hour. All shelters withstood the typhoon winds without any damage, giving confidence and encouragement to officials and beneficiaries alike. Based on this success the scheme was expanded and the United Nations Development Programme (UNDP) technical assistance was sought to help disseminate the technology throughout the country. The United Nations Disaster Relief Organisation (UNDRO) also participated in the project as a co-operating agency. The project was not simply concerned with the design and construction of the houses but also sought to develop and promote the value of self-reliance among the beneficiaries and the individuals' communities. For this reason it was decided to maximise the participation and commitment of these groups in the construction of their houses through community development and education.

To date more than 24,000 such core shelters have been completed and none have been destroyed in subsequent typhoons. The programme is continuing throughout the Philippines. Provision of the actual dwellings is but one of the achievements of the project. An overall improvement in the lives of the families has been achieved through the community development and income generation programmes. Families have higher incomes, are healthier, better educated and are more self-reliant as they take greater responsibility for their lives where previously they relied on government assistance.

The project is set to continue well into the future, as a proven success it receives wide ranging political support. A simple 'Do it Yourself Manual' for Core Shelter Construction has been prepared. Further checks on quality control, as well as a complete evaluation and documentation of the project, are underway.

There are four key innovative aspects to the project. Firstly, the typhoon-resistant design itself, secondly, the modular approach to construction, thirdly, the role of community involvement and participation, and finally the emphasis upon quality control and training in the project.

## Design and construction of the core shelter

The key innovative aspect of the design is the strengthening of the structure to enable it to withstand typhoons. Building research shows that the performance of a structure in high winds depends on the weight and design of the roof and how it is attached to the frame. Anchorage, tying the roof to the ground through cement footings, increases wind resistance, as does a four sided roof design strengthened by roof trusses. Roofing materials should be strong, shatter resistant and of medium weight. The core shelter technical design therefore incorporates the following typhoon resistant features :

1 Anchorage, tying the roof to the ground through cement footings to achieve durability.
2 A four sided roof design strengthened by roof trusses.
3 Use of extra bracing and anchoring on walls and ceilings to ensure stability.

An essential aspect of the design is that it should be easy to understand and build, thus making it acceptable to local people who can be trained in the simple methods necessary. The technology can also be easily transferred without the need for lengthy and complicated training courses.

Careful consideration is given to the siting of the core shelter houses - for practical reasons they are placed away from low lying areas which are regularly flooded, mountain slopes that are vulnerable to landslide, and on the immediate shoreline where they could be destroyed by tidal wave. In forested areas the houses are sited some 10-15 metres away from trees.

The core shelter itself is quite small, measuring 3 by 3.5 metres. Families typically build extensions around this core to increase the size of their accommodation, since average family size is 8 or 9 persons. In some cases although not all, these extensions are also built using the typhoon resistant technology. Where they are not, they run the risk of being flung to the four winds during typhoons, although their contents and fittings will have been moved into the core so that these are not lost. Wherever possible shelters are built in a plot of 80 square metres, enabling land around the shelter to be used for vegetable growing and other income generating activities. A typical shelter can be seen in Figure 8.1 below.

**Figure 8.1 A completed core shelter with extensions to rear and side**

*Design for improved wind resistance*

The shelter is rectangular in form with four wooden corner posts attached to concrete pedestals partially sunk in the ground and four wooden side posts situated midway on each wall, similarly attached. These firm footings, together with secure anchorage of the superstructure, help to ensure that the dwelling remains firm during typhoons.

The concrete footings for the corner posts each measure 0.5m x 0.5m x 0.15m and the corner post concrete pedestal is attached monolithically to the concrete footing. Each corner post pedestal itself measures 0.2m x 0.2m x 0.65m, and is made of Class A concrete mix, with reinforcing steel bars. The wooden corner posts (each 2.6m long) are attached to the footing pedestals by means of a steel post strap, partially embedded in the concrete corner post pedestal and extending above it, providing a site to fix the wooden corner post by means of nuts and bolts.

The side post concrete pedestals provide the concrete base for the four wooden side posts which are used to strengthen the roof girt. There are four of them, situated midway on the four sides of the house and they provide a firm support for the side posts. Each side post pedestal measures 0.15m x 0.25m x 0.4m and is made of Class B concrete. A U-shaped steel bar is inserted in the concrete and protrudes above it and is used to firmly anchor the floor girt of the dwelling

As already mentioned, corner footings and corner post pedestals are made using Class A concrete i.e. cement, sand and gravel used in a 1:2:4 mix. Side post concrete pedestals are of Class B mix i.e. 2:5:10, and for the flooring and concrete steps Class C concrete is used with a ratio of 1:3:6.

Special wind resistance features are included in the dwelling - these include wall and ceiling braces, anchor cleats and roof cross braces. The ceiling braces are made of 5cm x 13cm lumber and brace the two adjacent roof girts at 0.6m from the corner post. The brace length is 0.85m (See Figure 8.2).

The wall braces are made of 5cm x 8cm lumber and brace diagonally the corner post and the roof girt. The brace length is 0.90 metres and there are two braces fixed to each corner post, (as can be seen in Figure 8.3).

Anchor cleats are installed as a further means of holding the roof down and improving wind resistance. Anchor cleats are wooden blocks made of 5cm x 8cm lumber and they connect rafters or trusses to the roof girt. When connecting rafters to the roof girt they are 25.5cm long and are nailed only at one side. When connecting trusses to the roof girt they are 20cm long and are nailed on both sides.

The two roof trusses are constructed on the ground and then placed on the central portion of the roof girt, one metre apart. They are attached by wood cleats on both sides of the truss and nailed into position. A cross brace is installed to further strengthen rigidity in the roof and consists of two pieces of 5cm x 5cm lumber nailed in the form of a X, bracing together the two central roof trusses.

Rafters are then attached to the trusses and roof girt and additional steel strengthening is used at the corners where the hip rafters join the corner posts (see Figure 8.3). Figure 8.4 shows the positioning of the central trusses and wood cleats.

Figure 8.5 shows an interior view of a finished core shelter showing the ceiling and wall bracing and the anchor cleats connecting the roof truss to the roof girt.

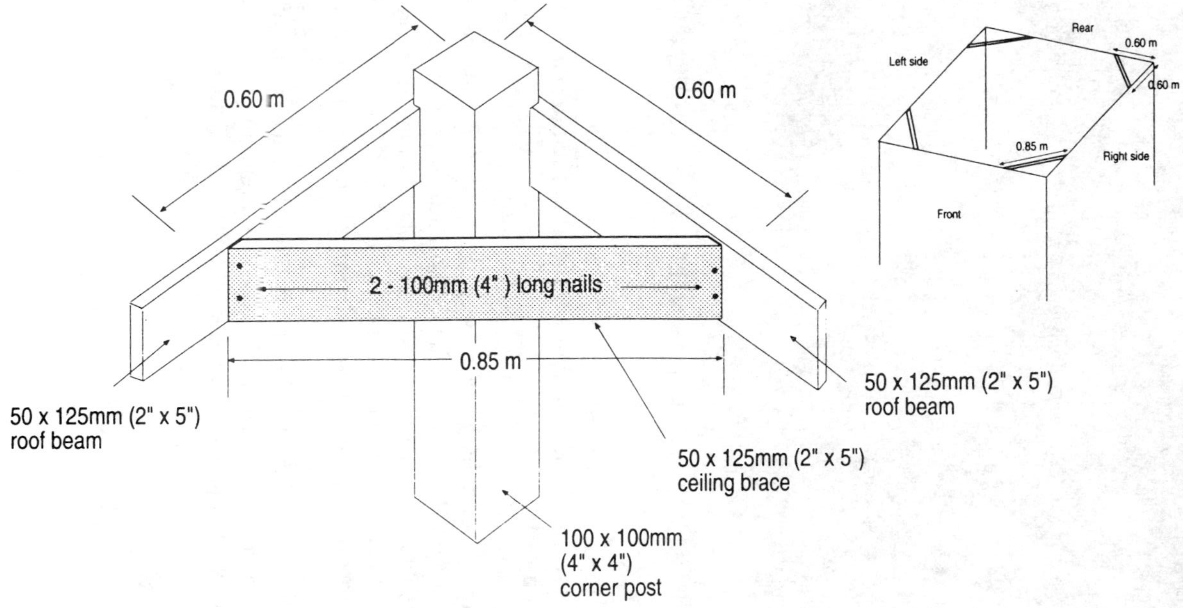


**Figure 8.2** **Showing positioning of ceiling brace**

**Figure 8.3 Wall bracing detail also showing metal ties joining corner post and hip rafter**

The completed frame of a dwelling can be seen in Figure 8.6, with the roofing work underway. Frequently the whole family will assist as general helpers. Roofing and wall cladding take only 1-2 days, the bulk of the work being in the construction of the frame. Each shelter requires a total of 47 man days for its construction but with a small team of people the work can be scheduled into ten working days.

A limitation of the core shelter design in the future will be the supply of wood which is required for the timber frame. Studies are currently being carried out into the use of other forms of lumber and in the cultivation of the necessary timber resources.

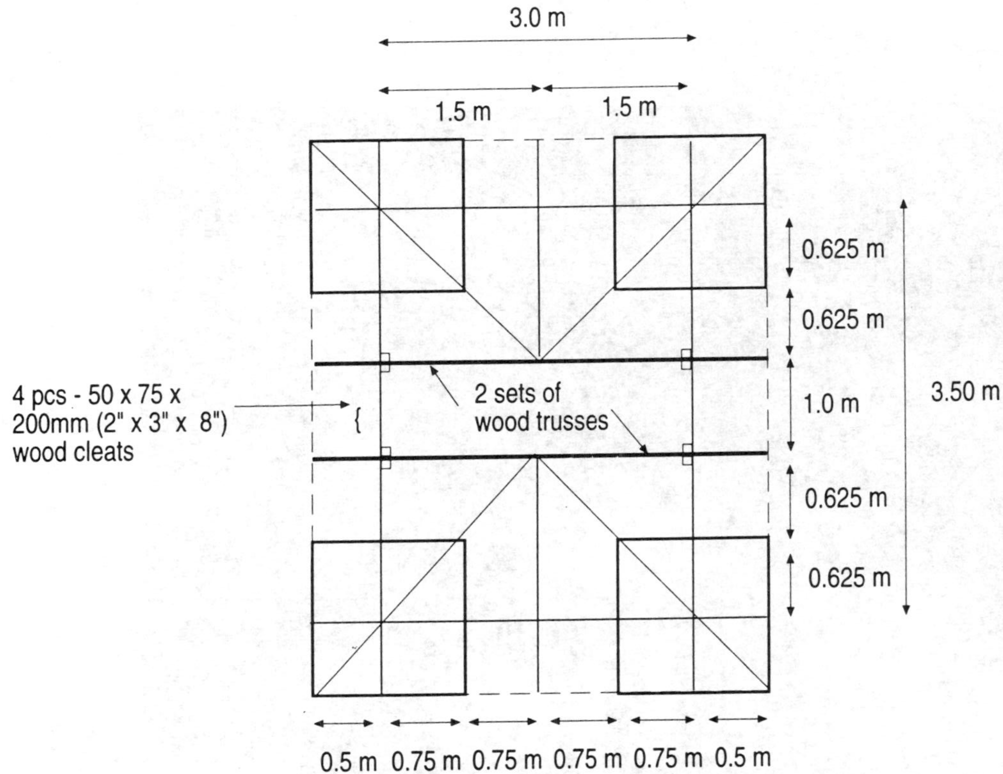


**Figure 8.4 Positioning of central trusses, cleats and rafters**

**Figure 8.5 Interior view of completed core shelter showing corner bracing**

**Figure 8.6 Wooden frame of core shelter, with roofing underway**

*Modular upgrading*

In addition to the basic core shelter dwelling, designs are available for three add-on modules, or options, for upgrading the dwelling. This enables each family to have a basic core shelter which is typhoon-resistant, which they can then upgrade at a later date, according to their resources and needs. Costs are kept down by using cheap, locally available materials for roofing, walling and flooring, since these units are not essential to the wind resistance of the dwelling. These modules are described below. Module A is the basic dwelling, B, C and D are the upgrading options.

| | |
|---|---|
| Module A | Basic core shelter unit consisting of foundations, wooden posts and frames, roof frames and trusses, gravel fill for toilet and flooring, and with indigenous local materials for all roofing, walls, doors and windows. |
| Module B | Upgrading and improvement through provision of Cement Hollow Block (CHB) walls and concrete slabs for main floor and toilet. |
| Module C | Provision of door and window panels and interior finishing. |
| Module D | Provision of galvanised iron corrugated roofing sheets, gutters, down pipes, etc.. |

This modular design incorporates all the features of modern low cost housing without imposing it upon the client and also enhances the flexibility of the project for different types of co-financing schemes. Above all it maximises the possibilities for client participation by enabling the shelter design to be adapted to available resources - salvaged materials etc. as well as by the use of indigenous roofing and walling materials.

## Community organisation and participation

The community development programme which runs alongside the housing programme is a key aspect of the project, since the whole project aims to improve not only people's shelter conditions but also other aspects of their daily lives. A range of health, education and income generation programmes run alongside the core shelter construction. Participation of the beneficiaries in the programmes is considered crucial and the DSWD spends a great deal of time on the social preparation stage of a project, prior to construction starting, in order that a strong community structure is in place.

Involving the beneficiaries in the project has several advantages. Firstly, it ensures that the type of shelter provided is exactly what the beneficiary wants and that needs are not exaggerated. Secondly, it guarantees the development of human capabilities and helps to avoid the creation of dependency. Thirdly, it generates a sense of pride and worth in the project, which means the beneficiaries value and safeguard the houses to a greater extent than they would otherwise do. In addition, the provision of post-disaster shelter provides an opportunity to develop co-operative action and spirit.

### *Community involvement*

A viable community organisation is required in order to mobilise resources, monitor the implementation of the project at the grass roots level and achieve a cost effective management and control over the resources used in the project. If the project is to be successful, thorough preparation is needed to enlist the support of local officials and politicians.

The DSWD holds meetings with municipal officials, mayors, priests, NGOs and community representatives. A community assembly is held to discuss the project and explaining the practicalities of how it works. The responsibilities and commitments of those involved are explained and potential participants in the project are identified. Questionnaire surveys are carried out by DSWD staff and trained volunteers, of potential participants to establish their needs and what resources are available. The community itself establishes eligibility requirements at its first assembly, and actual families are selected at a second assembly when the results of the surveys are known.

At this second assembly persons willing to be trained as foremen are identified. These are normally people with some existing building experience.

*Neighbourhood Associations for Shelter Assistance*

Once beneficiaries have been identified for participation in the project, they are organised into groups of at least ten families called Neighbourhood Associations for Shelter Assistance (NASA). Regular meetings are held throughout the project, initially to work out their housing plan, and later to ensure that it is being carried out, ironing out any problems as they arise. Individual housing plans are worked out for the selected families, identifying the specific materials needed and supply sources. NASA also determine land and manpower requirements. The DSWD provides those materials which cannot be found by the families themselves or by NGOs and local government organisations in the area. Local governments normally make land available to those communities which have to be relocated. A group of completed core shelters can be seen in Figure 8.7. As can be seen, construction has been carried out on a planned basis with clearly defined roads and walkways.

**Figure 8.7 Core shelter dwellings within a planned setting**

The beneficiaries receive training in construction methods and go on to build one unit of shelter under the guidance and help of a trained foreman, the other units in the group are then built with only limited supervision. Generally one trained foreman is responsible for the construction of 50 core shelter units, in co-operation with the beneficiaries and an area social worker. The NASA is responsible for mobilising labour resources to build the shelter units, and for the safekeeping of the set of basic carpentry and masonry tools that is provided by DSWD for every five shelter units. It also organises the storage of construction materials and monitors their use in the project and reports on the progress of construction work. The NASAs are also responsible for establishing a Shelter Maintenance Fund, with contributions agreed upon by each member and for carrying out subsequent maintenance of the dwellings.

The experience of the project is that this has proved a highly workable and beneficial system of operation. It does however require a substantial commitment in terms of staff time at the beginning of each individual project, both from technical staff and social workers. The integrated nature of the project has resulted in a greater merging of the two disciplines and a merging of the two sets of skills, with social workers acquiring technical expertise and technical experts developing skills necessary to foster community co-operation and support.

## Training and quality control

The attention to both training and quality control have been of crucial importance in determining the success of this project.

### *Training*

Training is an essential part of the Core Shelter Assistance Project, both for the officials involved in it and for the beneficiaries. To effectively implement the project, two types of training are carried out, relating to both the organisational and technical aspects of the project.

#### *Organisational training*

This is a one and a half day training course for DSWD regional technical staff, provincial and city technical supervisors, DSWD social workers and other field workers. It focuses on the theoretical aspects of the project. Topics discussed include:

a. Aims and objectives of the project.
b. Criteria in the selection of beneficiaries.
c. Components of the project such as social preparation, Food for Work, the technical and financial aspects of house construction.
d. Policies and guide-lines for the project.
e. Planning and monitoring of the project.

*Technical training*

This three day training course is focused on the technical aspects of house construction, as well as the evaluation of the progress of the project based on the action plans made during the first level training. In this session one core shelter is built and this is then used as a demonstration house for that area. The construction of the model core shelter is demonstrated by the National Consultant to the Regional Technical Consultant, Provincial and City Supervisors, DSWD Project Implementors and Foremen. Figure 8.8 below shows Professor Gupta explaining the construction of the roof trusses.

The social preparation component and the technical assistance on house construction are crucial components of the Core Shelter Project and are included in both levels of training.

**Figure 8.8 Instruction in the technical aspects of core shelter construction**

*Quality Control*

The credibility of any new technology is of key importance in ensuring its acceptability to rural communities and the core shelter is no exception. If the technical specifications of the structure are not followed correctly or there is faulty construction, the wind resistance of the house is reduced and thus its susceptibility to typhoon damage. This has the effect of discrediting the project and great efforts have been made to ensure that not only do the beneficiaries understand how they have to build the houses but why they have to do so.

There is continuous monitoring of work in progress to ensure that this knowledge is being put into practise and that the houses are being correctly constructed. The temptation is always to cut corners where possible, to use two nails instead of three, smaller timber than that specified and vigilant monitoring is necessary to ensure that construction is as specified. A certificate of completion is only given when the house is fully and properly completed. There is no better publicity for the project than when the core shelters are the only houses that remain standing and substantially undamaged after a typhoon has struck. A common experience is that villagers without core shelters will go to those of their friends and neighbours for protection in the event of a typhoon evidence of the project's credibility, the technology is also transferring to other families in the community who are not beneficiaries of the project - families who can afford to build their own homes are doing so incorporating the wind resistance features in their construction.

## Financial aspects of the Core Shelter Project

With a limited budget the DSWD found it necessary to assemble a package of funding sources to meet the costs involved in the project. The original funding came from the international disaster relief received for the victims of the Sisang Typhoon, but further extension of the project has had to be carried out using other sources.

### *Funding*

Funding for the basic core shelter before any upgrading (i.e. Module A) comes from five sources - the beneficiary, local government, the DSWD (i.e. central government), a Food for Work contribution and finally from NGOs. The DSWD contribution is in the form of a top-up contribution, meeting the requirement for whatever cannot be found from the other sources.

The contribution of the beneficiary is determined at the Needs Assessment stage, early in the project. This includes the provision of land, the labour of the beneficiary and friends and relatives, plus any materials which he can obtain, (either salvaged from the damaged house or from other sources) plus a financial contribution if these are available.

A beneficiary's income may well be affected because of the time that he spends building the shelter and in this case partial Food for Work assistance may be given. Emphasis is placed upon maximising the input of the beneficiary, not only so that the limited funds of the DSWD can be most effectively and extensively used, but also so that the commitment and involvement of the beneficiary is maximised.

The local government is in a position to effectively address the land related problems, by providing resettlement areas for those who are landless and those high-risk communities who live by the sea-shore. The DSWD provides materials and technical assistance for the construction of concrete footings and wooden roof trusses.

*Standard Costs*

Standard costs for the core shelter have been drawn up by the DSWD technical staff. These are prepared on a national basis and amended locally to take account of local variations. At the beginning of a project a complete Housing Plan is drawn up, itemising the sources of all the costs involved in the project, and enabling the financial input of local authorities and non-governmental organisations to be calculated. The standard costs include every last nut and bolt, each piece of wood etc. under the main headings of labour, land, concrete footings, formworks, wood framing, toilet and bath and roofing and walling. Similar standard costs are also drawn up for the three upgrading modules. The standard inputs for the basic core shelter (i.e. Module A without any upgrading) are 10 days labour and 12 square metres of land plus materials. Standard costs for a core shelter are set out in Table 8.1 below.

**Table 8.1**
**Standard costs for a core shelter (1991 prices)**

| ITEM | COST ($) |
|---|---|
| Housing materials | 735 |
| Food for Work | 225 |
| Technical assistance | 40 |
| TOTAL COST PER SHELTER | 1,000 |

Labour and the roofing and walling materials are the contribution to be made by the beneficiary. A grant of P8,000 is made to cover the cost of other materials for the beneficiary. The major component of material costs is the wood for the framing (55 per cent) with the foundation works (17 per cent), and the roofing and walling materials (18 per cent), the two remaining major cost items. The resettlement sites or individual home plots are provided either by local government or the NGOs.

Financial assistance has been received from the UNDP to the value of $175,000, subject to the availability of counterpart funding. This funding is used in training to improve the technical skills of foremen and beneficiaries.

*Cost Control*

A close check is kept upon the materials used in the construction of the core shelters, and considerable attention given to the best possible ways of delivering financial assistance. Rather than handing over a cash grant, a coupon scheme has been developed, whereby the beneficiary is given coupons for the purchase of materials from accredited suppliers, who are then paid by the DSWD upon receipt of proof of delivery. The beneficiary is given the coupons only after his Housing Plan and his entitlement to assistance is clearly established.

*Income Generation*

The importance of providing poor families with a means of earning an income is an essential part of the DSWD integrated development programme. As part of the Core Shelter Project beneficiaries of the housing programme are enabled to earn additional income,resulting in all round improvements in the living standards of the community.

The minimum plot size of 80 square metres insisted upon for all resettlement sites, gives families an opportunity to develop small gardening enterprises, capable of supporting self-sufficiency as well as generating a marketable surplus. The process of learning how to build the core shelter teaches building and in addition carpentry skills and a range of training courses are offered, covering a variety of other skills. Many of these generate small enterprises, such as fish-farming, pineapple cultivation, mushroom cultivation, basket work and the production of building materials such as nipa wall panels and are run by the women and children of the family as a supplement to the main family income.

Figure 8.9 below shows a fishpond within the core shelter plot, providing protein rich food for the family and a surplus for sale.

## Conclusion

The effects of natural disasters or 'Acts of God' vary according to the vulnerability of the person experiencing that disaster. Inevitably it is the poorest, those least able to prepare themselves adequately, who suffer most. This is certainly the experience in the Philippines.

The core shelter approach to disaster mitigation seeks to treat the causes of the problem as well as the symptoms, with its dual approach of community development alongside technical innovation. The vulnerability of the poorest people is reduced through training and income generation programmes, and the development of community based self-help and self-reliance, while the technical innovations in design limit the amount of destruction. This dual approach is bringing significant and long-lasting change to the lives of many thousands of people. The lessons shown here are widely applicable throughout the developing world where natural disaster and extreme poverty so frequently coincide.

**Figure 8.9 A fishpond within a core shelter plot**

**Further Reading**


Diacon, D., (1993). *A Practical Guide to the Construction of Low Cost Typhoon-Resistant Housing - The DSWD Core Shelter Project of the Philippines*. Building and Social Housing Foundation: Coalville.

Diacon, D., (1992). Typhoon Resistant Housing in the Philippines: The Core Shelter Project, *Disasters* Vol. 16, No. 3.

Gupta, S.P., (1991). 'Core Shelter Assistance Project for Typhoon Victims in the Philippines', in *Proceedings of the International Conference on Housing Through Support*, Building and Social Housing Foundation: Coalville, pp.35-39.

# 9 Building bridges to reduce risk:

## The Caqueta ravine, Lima, Peru

*David Sanderson*

### Introduction

The project 'Reducing Risk In Vulnerable Communities, Lima, Peru. Phase One: Caqueta Pilot Project' grew from a workshop: '*Rebuilding Communities in Lima*', held in January 1995. The workshop was organised by Lima's Engineering University, in conjunction with Oxford Brookes University, UK, and Massachusetts Institute of Technology, USA. The workshop was aimed at identifying realistic and affordable community development interventions in Lima's Caqueta district. During the workshop two key points became apparent:

1 That primary community problems could be viewed in terms of vulnerability, in particular to the threat of natural hazards. Furthermore that vulnerability to natural hazards was linked to, and increased greatly by, social, economic and political issues.
2 A fundamental problem was the lack of contact and ensuing mistrust between key actors: local community, the local municipalities, the national governmental disasters authority Defensa Civil, and to a lesser extent NGOs. This lack of good linkages between key groups was seen to be a major hindrance to sustained risk reduction.

These two points became the foundation for the initiative. The resulting project comprised two components: a risk evaluation of Caqueta's commercial and residential areas, and the Workshop 'Caqueta, Lugar Para Vivir Sin Riesgos' (Caqueta, Place To Live Without Risk). The risk evaluation focused on information gathering for hazard and vulnerability identification from communities, research organisations and government authorities. In addition meetings were held with institutions and individuals to ensure their participation in the forthcoming workshop. As well as undertaking participatory rapid appraisal (PRA) and desk top research, physical mapping of Caqueta, using satellite imagery and observation, was undetaken by IPADEL and the use of GIS software, as well as initial information gathering on organisational structures. The findings produced information regarding building height, age, materials used and the likelihood of collapse in the event of earthquake. At the end of the two weeks findings were presented to invited guests for review and feedback at the offices of Intermediate Technology in Lima. Guests attending included those from Municipalities, Defensa Civil, NGOs and academics.

The findings from the risk evaluation and, crucially, the relationships built up between key actors, formed the basis for the three day workshop held three months after the evaluation. At the workshop working groups comprised of members of each of the stakeholder groups identified key vulnerabilities and capacities, and possible initiatives aimed at reducing risk of natural disaster (notably landslide, fire and earthquake). Over thirty representatives attended the workshop, drawn from the local municipalities, traders and residents associations, local NGOs, the fire services, and international NGOs. The workshop comprised the presentation of key problems and the development of realistic, affordable initiatives in working groups. By the end of the workshop a number of initiatives were developed for risk reduction in Caqueta. These included training for fire awareness, ravine improvement through lobbying and information interchange, NGO awareness raising of disaster management, community level environmental improvement and bridge upgrading in the ravine area.

**Methodology**

Underpinning the project was the adherence to action planning[1], or action-oriented research. Action planning methodology states that rapidly gathered information, which may be incomplete and lack refinement can, if accurate, be used in the formulation of immediate projects where results can be seen. Such projects may be small or large scale, or may lead onto larger initiatives; but the point is that the research stage is the short period of a project, and is oriented towards what is possible; it is implementation that is the key part of action planning. Action planning appears as a response to the many large scale initiative methodologies for grand projects which have demanded intense detail and seemingly infinite amounts of technical and social data before action can begin: many such projects have failed when the budget, political and individual interest is exhausted in the research stage alone.

The situation in Caqueta is a good example of such failed initiatives: the risk evaluation uncovered grand projects to relocate entire communities, cover the ravine and implement city master plans which, however good and borne of genuine concern, have not happened because of the lack of political will, cash and/or the time for implementation. Hence, whilst the results of action planning may sometimes be criticised as incomplete and sometimes obvious, findings are being made, statements produced and action - real, tangible results - being implemented.

The project also relied methodologically on the pressure and release, or 'crunch', model and the need to increase capacities and reduce vulnerabilities.

*The crunch model*[2]

The crunch model, widely used in disaster management, states that risk (or disaster) is the product of vulnerability meeting a given hazard. Vulnerabilities may be social, economic, cultural, organisational or political, whilst natural hazards include earthquake, flood, landslide, volcano and fire. The diagram illustrates the relationship between hazard and vulnerability and allows for the identification of the components of these two elements: for hazards severity, duration, frequency, location; for vulnerabilities those mentioned above (see figure 9.1).

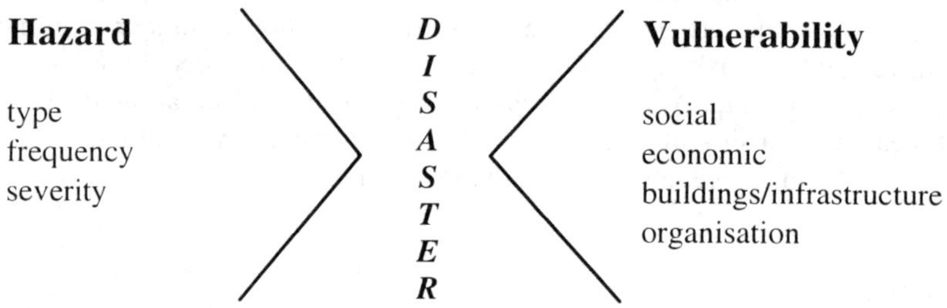


disaster = hazard + vulnerability

However it is also key to recognise **capacities**, hence the aim is also to

**reduce vulnerabilities and increase capacities.**

**Figure 9.1 The pressure and release (crunch) model**

*Capacities and vulnerabilities*[3]

The crunch model serves to articulate vulnerabilities (negative aspects causing disaster) but traditionally does not include capacities - positive aspects of a given situation which may, when mobilised, serve to reduce risk by reducing vulnerability. Reducing the risk to natural hazards can be described in terms of reducing vulnerability and increasing capacity. At first sight such a statement appears simplistic yet, within the complexities of risk reduction within the minefield of political, economic and social factors, dictated by interest groups and each with thier individual agendas, agreement of the lowest common denominator for reducing risk provides the framework for movement forward.

This definition for action is useful for two other key reasons:

1 It is readily comprehensible by all. So often the notion of risk reduction is a stumbling block - a difficult to grasp concept seemingly unlinked to everyday life. To provide an easily understood framework usable at community, NGO, governmental and intergovernmental level begins a dialogue of commonality between key players

2 The definition forces a recognition of what is *good* in a given situation, not only on what is bad. Capacity is a positive statement recognising existing measures and mechanisms already employed for reducing risk.

Hence the aim of the project was to identify hazards, vulnerabilities and capacities, with a view to ultimately identifying ways of reducing vulnerabilities and increasing capacities.

## Research methods

During the risk evaluation four research tools were employed:

*Collection and review of existing research*

The OCDS/IPADEL research team was anxious not to replicate any existing work carried out; it was also aware of its own limitations of time and resources and was relying therefore on previous studies for technical information, e.g., earthquake in Lima, soil conditions in the ravine. To these ends technical information and background information were gained from the 'Rebuilding Communities in Lima' workshop, January, 1995; previous work undertaken by Intermediate Technology, including in particular that of Andrew Maskrey and La Red; meetings with Defensa Civil; and earthquake vulnerability information from CISMID (the Japan-Peru Japan-Peru Centre for Earthquake Research Centre for Earthquake Research) at the Engineering University.

*Observation*

The purpose of the observation teams was to build up hazard maps of Caqueta relating to buildings and infrastructure. Data collected included approximate building age, state or condition, number of floors and construction materials. The observational research was carried out in two stages:

- prior to the risk evaluation by an IPADEL team who surveyed three sections of Caqueta, placing the findings onto computer for hazard map production.
- during the risk evaluation, when two sample areas were observed, the findings combined with house to house interviews collecting socio-economic information to form a 'picture' of small communities.

*Questionnaires at household and organisational level*

Two sets of questionnaires were used to gather information. These were:

- Organisational capacity. IPADEL staff conducted a questionnaire with local associations (market traders and social organisations) in Caqueta to provide background information on organisational capacity. Information gathered related to size of organisation, activities, breakdown of members by age and sex, frequency of meeting. Information also gathered related to perception of risk by ranking.
- Households. Brief questionnaires relating to social and economic data were conducted with households in each of the two typologies relating to dwelling. Information gathered related to number, age and sex of household members, income earning activities and approximate monthly incomes of family members.

In addition to the gathered data, life stories were gathered of a sample of households in order to provide qualitative data. A check list was provided for the interviewer with questions relating to time living in household, reason for choice of location, and a recounting of any major disasters which affected each family's or community's livelihood.

*PRA meetings*

A series of meetings were held with market and housing association representatives to gather data using PRA methods. These included:

- Community mapping. On large sheets of paper participants were encouraged to draw their area (usually several blocks) and to map key buildings, eg cinema, storage depository in markets, church, etc.

- Time line development. A time line, usually from either 1970 or 1980 (depending on the age of the participants) until the present day was drawn. Participants indicated along the line significant good and bad events of key importance; the former may include the finishing of roads or the installation of services; the latter landslide, tremors or significant fires.
- Hazard mapping. The map was returned to participants who were asked to indicate on the map where these hazards had occured.
- Development of disasters matrix. Identified disasters were placed on a matrix, with the headings of disaster, causes and possible solutions. Participants were also asked to state which was most important, second most important etc.
- Discussion of results. Finally results were discussed in the group, mostly around the matrix; many participants shared their own visions of how the identified situations could be improved.

**Figure 9.2 PRA meeting at Association Dos de Mayo**

**Natural hazards in Lima**

The ravine area suffers from earthquakes and landslides along the Rimac valley which transects the city. In addition, poorly enforced building and planning codes, high densities and rapid urbanisation, increase fire risk in the area.

*Earthquake risk in Lima*[4]

Peru lies at the junction of the Nazca and South Americana Plates. Of all seismic activity in Peru, Lima ranks as the area with the second highest level of activity[5]. Since 1586 nine major earthquakes have been recorded in Lima, the last one being October 3, 1974, when 78 people died and a reported 2,550 were injured. In particular Lima is vulnerable to earthquake damage due to soil conditions - it is built on a desert strip and the flat land is comprised of boulders, gravel and sand. In his book *Community Disaster Mitigation* (Oxfam, UK, 1987), Andrew Maskrey identifies three areas within Lima according to vulnerability, of which the ravine falls into the third category:

- middle and high income areas - low vulnerability.
- Pueblos Jovenes, e.g. Villa el Salvador - low vulnerability.
- inner city - high vulnerability.

In 1983 the National Institute for Urban Development (INADUR) undertook studies to prepare a Seismic Protection Plan for Lima Metropolitan Area for Defensa Civil. The Seismic Plan however was never implemented (Maskrey, 1987), studies did find however that large amounts of housing were highly vulnerable for two key reasons:

- poor resistance of original structures and materials used.
- poor maintenance, combined with intensive use of buildings through overcrowding.

The study identified two housing types in the old centre: those over 50 years old - older, larger houses built with heavy roofs, adobe walls and with wooden joists were traditionally highly resistant to earthquake. However their resistance has been reduced through over use and neglect. The key reason for deterioration is overcrowding, caused by the subletting of rooms within such buildings by low income migrants. The second housing type is houses 30-40 years old - smaller, one storey adobe, built with lighter roofs, and less vulnerable than their larger neighbours because of their small size and lighter weight.

Defensa Civil's study of one area of Lima, Lima-Cercado, found that of the 73,091 houses in the area, approximately 25 per cent (18,087) were under risk of collapse. Another key problem identified was the presence of ambulantes which, in the event of a disaster, would not only themselves be highly vulnerable, but would also restrict rescue services access.

A further study by CISMID (Rios, 1991) in Barrios Altos, the same district of Lima-Cercado, predicted that of the 1,964 houses surveyed, 58 per cent (1,137) were considered vulnerable to collapse, 13 per cent (258) to severe damage and 29 per cent (571) to moderate to light damage. Of the population of 9,814, a moderate earthquake would therefore cause 100 deaths, 1,000 injuries and 7,000 persons to be displaced.

*Landslide and the ravine*

North of Lima the Rimac Valley, which forms the Caqueta ravine, is subject to severe mudslides, or *huaicos*. The huaicos occur from December - February each year, causing floods within the Rimac Valley floodplains as well as rapidly moving torrents of water within the valley. The effects of huaicos are felt in Chaclacayo and Lurigancho-Chosica, peripheral districts of Lima; also Lima's water supply is affected, as are transportation routes in and out of the city. In addition farmland is affected. The effect on the River Rimac within Lima (including that flowing through Caqueta), is of torrents of water flowing in January and February each year. At that time the river may rise by up to five metres. During PRA exercises perceptions of the huaico, given by ravine edge residents in Caqueta, included:

- 'it has no effect on us since the ravine is so deep'
- 'it is a good thing since it serves to remove the rubbish and sewerage gathered on the ravine edges'

However the annual huaico does serve to further erode the ravine, deepening it and ultimately widening it by causing further landslide along its deep banks. Some evidence of the rate of erosion may be gained from conversations with local residents: older people remember being able to wade across the ravine in the early 1950s (although this is uncorroborated). Other indicators by local residents state that the ravine is widening by about one meter per year. Buildings now right on the edge of the ravine, less than forty years old, which were hitherto on stable land bear witness to rapid erosion of banks.

Juvenil Medina Rengifo, disaster specialist at Intermediate Technology, carried out a study of the geological conditions of the ravine area: it is his belief that the ravine, and other similar ones within Lima, are the result of geological faults, thus the erosion caused by the Rimac river, and especially the coming of the huaico, serve to exacerbate an already existing problem.

Due to its location, the problems of the ravine have a high profile within Lima. Several initiatives at governmental level have been discussed for solving the problem. Amongst these are:

- Covering the ravine: during the period of President Belaunde's administration in the mid 1980s, a project was developed to cover the ravine with a lightweight structure, on which industrial enterprises would operate. The plan also included relocation of the ravine dwellers to improve the area. The initiative did not come to fruition.
- Relocation to Ventanilla: it was the view of the Chief Adviser to the head of Defensa Civil that all of the area neighbouring the ravine, and those dwelling on the former garbage site, should be relocated to Ventanilla, a government owned relocation area far from Caqueta in the north cone of Lima. The ravine area would be zoned as parkland, all buildings and infrastructure would be removed and the area replanted.
- Creation of the Rimac River Authority: legislation currently with Parliament for debate proposes the creation of a rivers authority for the Rimac River. Such an authority would provide one body of power responsible for the Rimac, including the ravine and those dwelling on it.

Little has changed so far however, to improve conditions. Whilst ultimately solutions will doubtless rely on some degree of government intervention, discussions with local communities, experts and political bodies, have revealed that the vulnerabilities of the governmental approach to date include:

- Reliance on large scale technical solutions. Solutions to date such as concreting the ravine banks would be technically complex and hugely expensive.
- Lack of co-ordination at political level. The Rimac ravine transects several municipal districts: a co-ordinated strategy would be hard to achieve with different interests, points of view, etc. to be reconciled.
- Lack of political will. Forced relocation on a large scale would be politically unpopular amongst local communities.
- A desire to solve all problems in one macro level initiative. The top down approach of one solution has to date proven impossible. A cocktail of small scale initiatives with differing time frames might prove more realistic.
- Limited resources. Large scale relocation (including new housing) and urban improvement (demolition of vulnerable housing and reuse) is very costly.
- Lack of community consultation. Communities themselves are aware of the problems, but choose to live in vulnerable areas in order to be close to work. The research found that many families

**Figure 9.3a & b The Caqueta ravine**

had moved to dwellings on the edge and intend to sell on their properties.

## Risk evaluation findings

The following information presents the key findings of data gathering within the ravine area. The findings include a summary of hazard, vulnerability and capacity, and PRA findings. PRA was undertaken with residents of Association Dos de Mayo, whose borders extend to the ravine edge. One street along the ravine edge was also chosen for a more detailed survey of building condition, residents and their incomes. Data gathered was used as the basis for the following 'Caqueta, Place To Live Without Risk' workshop.

*Historical profile*

Community representatives were asked about particularly memorable natural disasters. Residents stated buildings fall into the ravine 'all the time'; children also fall in from the edges and from the bridge on a weekly basis (the sides in places have been removed for the tipping of garbage). Ironically the garbage often saves lives by providing a 'cushion' for those falling in.

**Table 9.1**
**Major events in Caqueta ravine since 1968 (from PRA studies)**

| DATE | EVENT |
|---|---|
| 1968 | tremor & landslide: 4 houses and 2 families effected |
| 1972 | law passed to move local chemical producing factory to Ventanilla |
| 1978 | landslide: 2 houses lost and 1 family effected |
| 1983 | landslide: collapse of coliseo |
| 1984 | landslide: collapse of wall |

**Table 9.2**

**Key problems and possible solutions in Caqueta ravine, as identified by community representatives (from PRA studies)**

| DANGERS | PROBLEMS | 'SOLUTIONS' |
|---|---|---|
| Landslide of houses into the ravine. | Ravine is very steep. Broken sewage pipes leak into the soil. Erosion of river undercuts ravine. | Raising of the height of the river. Construction of a covering over ravine. |
| Bridge is in danger of collapse. | Foundations are exposed. | Repair and replace wood en walkway with concrete walkway. |
| Pollution of river Rimac. | Dumping of garbage and sewage. | |
| Pollution of air by local factories. | Health of community, especially eye, nose and ear problems. | Relocation of factory away from residential zone. |

**Hazard**

**earthquake**
infrequent but very damaging
loss of livelihood and lives

**tremor**
more frequent but less damaging
structural damage

**fire**
very frequent
mainly loss of livelihood (eg markets)

**landslide**
very frequent in ravine
mainly loss of investments (housing)

*DISASTER*

**Vulnerability**

dwellings on ravine edge
poor quality land: unstable, eroding
illegal land (no ownership)
no information on disasters
old rescue services equipment
poor building maintenance and construction
no enforcement of building codes
co-ordination at institutional level: ravine covers several municipalities
fear of co-ordinating institutions
forced relocation: loss of investments and access to jobs
low or diminishing value of properties

*but*

**Capacity**

value in land and buildings: market of buying and selling
community investment: buildings and infrastructure
location near to employment opportunities (in the city centre)
awareness of issues and attempts at improvement
strong community organisation,: elected leaders, ravine included in legally owned Dos de Mayo area
aspirations: self improvement
local skills and abilities
commitment of rescue services
commitment to improvement by authorities
high public profile: ravine frequently on television when fatalities occur

**Figure 9.4 Summary of key hazards, vulnerabilities and capacities in the ravine**

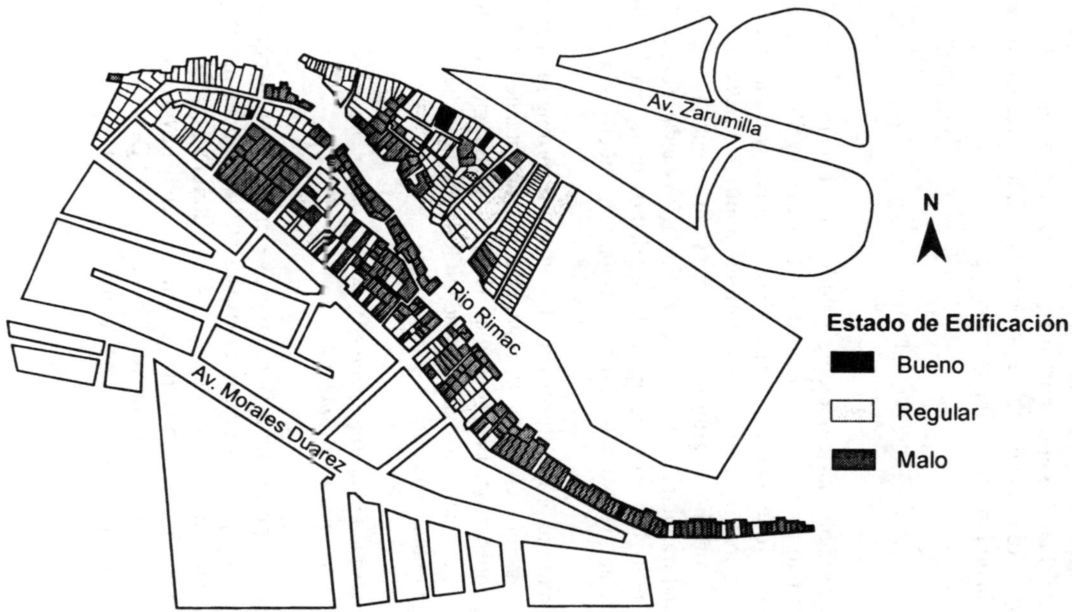


**Figure 9.5 Building condition, Caqueta Ravine (original in colour)**
Maps were also produced for age of buildings, number of storeys and construction materials.

*Household Surveys*

These found that households on the ravine edge had access to incomes; total monthly household incomes ranged from 200-1,000 Soles. Residents had been on the edge from between 2 and 50 years. Despite the dwellings being officially illegal, (building is prohibited within 25m of the ravine edge), all had bought their properties, and most would be anticipating selling them on. Hence despite the clear risk associated with living on the edge of the ravine, property retains an exchange value. Families are continuing to buy and sell as well as consolidate their property. Additionally there appears a high diversity of employment, including formal and informal employment types.

**Next steps**

There is no affordable, all embracing solution, to reducing vulnerability in the ravine. Past initiatives bear witness to this. However whilst stalemate continues and little is done, (due to lack of resources, political will and the exclusion of actors in decision making), other interventions to *mitigate* risk are possible. As a result of the workshop, groups are better in touch (especially between communities and municipality); also the profile of risk awareness and the need to reduce vulnerability has been highlighted. Workshop participants from the ravine dwellers' association also talked about organising their own risk awareness meetings.

As Phase Two of the project further interventions, regarding ravine improvement through lobbying and community organisation, are planned. Of the seven initiatives developed from the workshop, (including strengthening fire services, improving fire awareness in markets and NGO training), the two key projects directly relating to the ravine are:

1. *Ravine Environmental Improvement Programme; Reducing risk through participatory urban improvement*

   The aim of the programme is to improve environmental conditions in the ravine area of Caqueta. An essential aspect of the eighteen month project is to inform and mobilize all interest parties - at community, NGO, municipality and government levels - towards common agreement for action. The project comprises two stages: Mobilization for Action, and Dissemination. Stage One comprises three related activities, aimed at informing interest parties and galvanizing action. Hence:

i) Mobilization for Action

- Mobilization at Community and Municipality Level. Through a series of community and municipality level workshops, participants will explore options for improvement and develop strategies for their implementation. Hence *continuing improvements at community level will be a feature for this activity.*
- Institution building. Governmental, municipal and NGO level organisations will be invited to awareness raising seminars/workshops, as well as being provided with information on the ravine, opportunities, problems, etc. In addition different institutions will be invited to work together in the formation of a *ravine improvement working group.*
- Feasibility studies. Technical, economic, legislative and social options will be explored by local and international experts for environmental improvement.
- Communication. Throughout the project information from one activity will be made available to the other through a quarterly news-sheet.

ii) Dissemination

The outputs from each of the three activities will be disseminated widely as options for improvement. Information will be presented at Governmental, NGO and municipality levels with a view to mobilizing action for change, and for securing national funding for implementing change.

2. *Ravine Environmental Improvement Programme; Bridge Improvement Party for Community Mobilization and Risk Reduction*

Funding for this project has already been provided by the British Embassy in Lima. The project is due for completion in September 1996. The project is improving two pedestrian bridges crossing the densest area of the Caqueta ravine, through community mobilization and self help.

The principle benefits to the community are to:

- prevent accidents and death, especially of small children
- increase safety for other pedestrians
- mobilize communities for greater self help and taking responsibility for their own neighbourhoods

The idea is to improve the bridge through a bridge improvement party, wherein organised groups will provide new solid, irremovable bridge sides.

After fixing the sides others, principally women and children, will paint murals on the sides and in the evening there will be a party.

## Towards increasing capacity and reducing vulnerability

Some of the best initiatives however require no funding, rather a change in values and perceptions of the key actors. In concluding the following five points, evolved during the project, stand as 'statements of belief' for reducing vulnerability and increasing capacity.

1 *Effective mitigation measures include participation in decision making at all levels.* A key vulnerability lay in the lack of trust between different key actors: municipality, NGOs, Defensa Civil, community groups. Each appears to perceive the other as having a different agenda to their own. Communication is often limited to directives, ultimatums or resistance.

2 *Perception of acceptable risk, however extreme, is outweighed by livelihood considerations.* The ravine edge dwellers are only too aware of the risks that they face, however the proximity of the ravine to the city centre, where income earning is possible, and the lack of affordable shelter elsewhere are irreconcilable factors.

3 *A key capacity is people: aspirations for betterment, skills and abilities.* The existence of strong community organisation, the upgrading and maintenance of shelters, is testimony to the ability of individuals and communities to cope with such extreme conditions. All too often this vital capacity is overlooked, especially by governmental authorities who can remain oblivious to the capabilities of communities.

4 M*acro scale solutions may ignore micro scale community capacity.* The 'solutions' of the governmental authorities lay exclusively in macro scale initiatives, forced relocation of entire communities - measures for which gaining political and economic support would be titanic. Whilst aiming at such large one off gestures, smaller, mitigating activities are being ignored, which may not offer solutions but would improve conditions.

5 E*ven the most vulnerable areas have economic value.* A key finding of capacity was in the existence of a property market on the ravine edge, even amongst some of the most vulnerable dwellings. Several of those interviewed had lived in their houses for two or three years. Many saw their dwellings as 'stepping stones', where families would dwell for a few years before moving on to better housing.

## Notes

1 Action planning is described in *Housing Without Houses* by Nabeel Hamdi, Van Nostrand Reinhold, New York, 1992; in addition *Making Microplans* by Nabeel Hamdi and Reinhard Goethert, IT Publications, London, 1988.
2 The crunch model, or pressure and release model is used in the United Nations DMTP manual *Vulnerability and Risk Assessment* by Coburn, Spence and Pomonis; in addition recent publications including *At Risk*, by Davis, Wisner, Blakie and Cannon, Routledge, 1994.
3 A clear explanation of capacities and vulnerabilities can be found in the chapter 'A framework for analyzing capacities and vulnerabilities' in *Rising form the Ashes; development strategies in times of disaster*, by Mary Anderson and Peter Woodrow, Westview Press (Boulder, Colorado) and UNESCO Press (Paris, France), 1989, pp.9-25.
4 Earthquake risk in Lima is summarized in the paper 'Seismic safety of the Lima metropolitan area', by Jorge E Alva Hurtado, Professor of Civil Engineering, CISMID, presented in the publication *Issues on Urban Earthquake Risk*, B.E. Tucker *et al*, 1994, Kluwer Academic Publishers, The Netherlands, pp.251-264.
5 Lima has a Modified Mercalli Intensity level of X, only superseded by XI to the south of the country.

# Bibliography


Alexander, D. (1993). *Natural Disasters*, UCL Press: London.

Alexander, D. (1995). *Natural Disasters*, UCL Press: London.

Arsovski, T. (1985). 'Reconstruction of the Centre of Skopje.' *Ekilizia Popolare*, Vol. 187, pp.26-37.

Arsovski, T. (1988). 'Post-Earthquake Urban Zoning: Trends in the Reconstruction of the Town, Revitalisation of the Historic Heritage and Damaged Area.' *1st International Seminar on Modern Principles in Conservation and Restoration of Urban and Rural Cultural Heritage in Seismic-Prone Regions*, pp.83-96.

Arsovski, T. (1989). *Skopje: Urban Planning and Architecture: 1945-1985*, Vol. 3, Sobranie na Grad Skopje: Skopje. In Macedonian.

Arsovski, T., and Taskovska-Arsova, N. (1988). *Urban History of Skopje From the 19th and the Geginning of the 20th Century*, Vol. 2, Sobranie na Grad Skopje: Skopje. In Macedonian.

Association of State Floodplain Managers. (1990). 'Challenges Ahead: Flood Loss Reduction Strategies for the 90s.' *14th Annual Conference of ASFPM*, Wisconsin.

Awotona, A. (1992). 'Approaches to post-war reconstruction and development: Lessons from Africa.' *Habitat International*, Vol. 16, No. 4, pp.79-98.

Barakat, S. (1993). *Rebuilding and resettlement, 9 years later; a case study of the contractor-built reconstruction in Yemen, following the 1982 Dhamar earthquake*, University of York (IoAAS): York.

Berke, P. (1995). *Natural Hazard Reduction and Sustainable Development: A Global Assessment*. Working Paper S95-02, Centre for Urban and Regional Studies, The University of North Carolina at Chapel Hill: Chapel Hill.

Blaikie, P., Cannon, T., *et al.* (1994). *At risk: natural hazard, people's vulnerability and disasters*, Routledge: London.
Blazevska, Z. (1982). *Contemporary Structual Housing Concept*, Unpublished Masters dissertation, Universitet Kiril i Metodij: Skopje. In Macedonian.
Blazevska, Z. (1983). 'Socio-Psychological Analysis of Residential Areas in Skopje.' *Zbornik na Arhitecktonskiot Fakultet vo Skopje*, Vol. 6, pp.121-150. In Macedonian.
Blong, R. (1992). 'Some Perspectives on Geological Hazards', in G. McCall, D. Laming, and S. Scott, (eds.), *Geohazards: Natural and Man-made*, Chapman & Hall: London, pp. 209-216.
Bobic, M., and Vujovic, S. (1985). *'Shelter: Residential Slums and Poverty.'* Filip Bisnjic: Beograd. In Serbo-Croatian.
BUET, BIDS, *et al.* (1992). *Multipurpose Cyclone Shelter Programme Inception; Interim and Draft Final Reports*, Government of the Peoples Republic of Bangladesh/United Nations Development Programme/World Bank.
Building Design. (1990). 'Architects Pack Prince Charles Ideas; UIA Plan for Crisis Homes Competition.' *Building Design*, No. 970, January 26.
Burgess, R.G. (1984). *In the Field*, Allen & Unwin: London.
Burton, I., Kates, R., *et al.* (1978). *The Environment as Hazard*, Oxford University Press: New York.
Burton, I., Kates, R.W., *et al.* (1993). *The Environment as Hazard*, Oxford University Press: London.
Casley, D.J., and Lury, D.A. (1981). *Data Collection in Developing Countries*, Clarendon Press: Oxford.
Chan, N.W. (1995). *A Contextual Analysis of Flood Hazard Management in Peninsular Malyasia*, PhD Thesis, Middlesex University: Enfield.
Choguill, C. (1994). 'Implementing Urban Development Projects: A Search for Criteria for Success.' *Third World Planning Review*, Vol. 16, No. 1, pp.25-39.
Chowdury, A.M.R., Bhuyia, A.U., *et al.* (1993). 'The Bangladesh Cyclone of 1991: Why so many people died.' *Disasters*, Vol. 17, No. 4.
Coburn, A., and Spence, R. (1992). *Earthquake Protection*, John Wiley and Sons: Chichester.
Cole, H.T. (1995). 'The Liberian Civil Servant's Plight, Low Remuneration-High Cost of Living: A Miracle to Survive.' *The New Democrat*, Vol. 2, No. 82.
Correia, F.N., Saraiva, M., *et al.* (1994). 'The Planning of Flood Alleviation Measures: Interface with the Public', in E. C. Penning-Rowsell and M. Fordham, (eds.), *Floods Across Europe: Flood Hazard Assessment, Modelling and Management*, Middlesex University Press: London, pp. 167-193.
Cuny, F. (1981). 'Disasters and the Small Dwelling: the State of Art', in I. Davis, (ed.) *Disasters and the Small Dwelling*, Pergamon: Oxford.
Cuny, F.C. (1983). *Disasters and Development*, Oxford University Press: Oxford.
Cuny, F.C., and Tanner, V. (1995). 'Working with communities to reduce levels of conflicts.' *Disaster Prevention and Management*, Vol. 4, No. 1.

Davis, I. (1975). 'Skopje Rebuilt - Reconstruction Following the 1963 Earthquake.' *Architectural Design*, Vol. 11, pp.660-663.

Davis, I. (1986). 'Lessons From Reconstruction After Natural Disasters for Cities Recovering From Bombing and Civil Strife.' *International Conference on Reconstruction of the War-Damaged Areas*, University of Tehran, Iran. Paper presented at conference.

Davis, I. (1987a). 'A Manual for the Implementation of Disaster Risk Reduction Measures.' *Open House International*, Vol. 12, No. 3, pp.5-15.

Davis, I. (1987b). 'Safe Shelter within Unsafe Cities: Disaster Vulnerability and Rapid Urbanisation.' *Open House International*, Vol. 12, No. 3, pp. 5-15.

Department of Humanitarian Affairs (DHA). (1993). 'Natural Disasters'. *DHA News,* Department of Humanitarian Affairs: Geneva, Jan/Feb, pp.26-27.

Devas, N., and Radoki, C. (1993). *Managing Fast Growing Cities: New Approaches to Urban Planning and Management in the Developing World.* Longman: Harlow.

Dorgan, A.C. (ed.) (1995). *Gale Country and World Ranking Reporter.* Gale Research Inc. - ITP: New York.

Doxiadis Associates, and Institute of Town Planning and Architecture. (1965). *Skopje Urban Plan: Housing*, Vol. 13, Zavod za Urbanizam i Arhitecktura: Skopje, Macedonia.

Faour, A. (1988). 'The War and People Displacement in Lebanon.' *Al-Moustakbel Al-Arabi*, Vol. 11, No. 114, pp. 95-120. Beirut, Arabic periodical.

Financial Times. (1995). *Finanacial Times Survey: Republic of Macedonia.* Financial Times: London, No. 32,721, 7 July, pp. 33-36.

Fister, P. (1988). 'The Influence of Siesmic Activities on Conservation, Rehabilitation and Preservation of Urban and Rural Architectural Heritage in Regional Parameters.' *1st International Seminar on Modern Principles in Conservation and Restoration of Urban and Rural Cultural Heritage in Seismic-Prone Regions*, Skopje, pp.39-45.

Flood Plan Coordination Organisation. (1992). *FAP-12, FCD/I Agricultural Study*. Final Report Vol. 1, Bangladesh Flood Action Plan, Government of the People's Republic of Bangladesh: Dhaka.

Foster, H.D. (1995). 'Disaster Mitigation: The Role of Resilience.' *Tri-Lateral Workshop on Natural Hazard Risk Assessment: Canada/USA/Mexico*, Merrickville, Ontario.

Galic, R. (1977). *Economic Evaluation of the Development Plans of Karpos, Topansko Pole, Aerodrom and Kapistec Housing Estates*. Unpublished report Arhitektonski Fakultet, Univerzitet Kiril i Metodij: Skopje. In Macedonian.

Golubovski, D. (1995). 'Two Places in Skopje: on Corner Houses and House Corners.' *Archin*, Vol. 1, pp.10-19. In Macedonian.

Green, C.H., van der Veen, A., *et al.* (1994). 'Vulnerability Refined: Analyzing Full Flood Impacts', in E. C. Penning-Rowsell and M. Fordham, (eds.), *Floods Across Europe: Flood Hazard Assessment, Modelling and Management*, Middlesex University Press: London, pp.32-68.

Green, R. (1995). 'Widening Horizons in Post-war Reconstruction.' *Disaster Prevention and Management: An International Journal*, Vol. 4, November 1, 1995.

Greene, M.R. (1987). 'Skopje, Yugoslavia: Seismic Concerns and Land Use Issues During the First Twenty Years of Reconstruction Following a Devastating Earthquake.' *Earthquake Spectra*, Vol. 1, pp.103-117.

Hakim, C. (1987). *Research Design: Strategies and Choices in the Design of Social Research*, Unwin Hyman: London.

Handmer, J.W. (1995). 'Issues emerging from natural hazard research and emergency managent.' *1995 Fenner Conference on the Environment: Risk and Uncertainty in Environmental Management*, Canberra.

Haque, C.E., and Blair, D. (1992). *Vulnerability to tropical cyclones: evidence from the April 1991 Cyclone in coastal Bangladesh*, Mimeo.

Hardoy, J., Mitlin, D., *et al.* (1992). *Environmental Problems in Third World Cities*, Earthscan: London.

Harphan, T., Lusty, T., *et al.* (1988). *In the Shadow of the City: Community Health and the Urban Poor*, Oxford University Press: Oxford.

Hershfield, A.F. *et al.* (1983). 'Fieldwork in Rural Areas', in M. Bulmer and D. Warwick, (eds.), *Social Research in Developing Countries: Surveys and Censuses in the Third World*, John Wiley: Chichester, pp.241-251.

Hewitt, K. (1983a). 'The idea of calamity in a technocratic age', in K. Hewitt, (ed.), *Interpretations of Calamity; from the viewpoint of human ecology*, Allen and Unwin.

Hewitt, K. (1983b). '*Interpretations of calamity.*' Allen and Unwin: Winchester.

Institute on Town Planning and Architecture. (1985). *City of Skopje Urban Plan: Amendments*, Vol. 7, Sobranie na Grad Skopje: Skopje. In Macedonian.

International Federation of Red Cross and Red Cross Societies. (1993). *World Disaster Report, 1993*, Martinus Nijhoff: Dordrecht.

Islam, N.K.M. (1996). *The impacts of flooding and methods of assessment in urban areas of Bangladesh*, PhD (in preparation), Middlesex University: Enfield.

Jeffery, S.E. (1980). 'The creation of vulnerability to natural disaster: Case studies from the Dominican Republic.' *Disasters*, Vol. 6, No. 1, pp.38-43.

Johnson, M.G. (1995). *Towards an Efficient Resettlement and Reconstruction Programme in Post-War Liberia*, Unpublished M.A. Dissertation, University of Newcastle: Newcastle upon Tyne, UK.

Jordanovski, K. (1993). *Skopje: Catastrophe - Reconstruction - Experience*, Matica Makedonska: Skopje. In Macedonian.

Kelly, C. (1995). 'Assessing Disaster Needs in Megacities: Perspectives from Developing Countries.' *Geojournal*, Vol. 37, No. 3, pp.381-385.

Kesseley, S.V. (1995). *In Liberian Civil War, 300,000 Persons Killed or Wounded.* The News Newspaper: Monrovia, Liberia.

Khalaf, S. (1989). *Besieged and Silenced; the Muted Anguish of the Lebanese People (Prospects for Lebanon)*, Centre for Lebanese Studies: Oxford.

Korobar, V.P. (1995). 'The Skopje Plan and the Crisis in the Urbanism of the Sixties.' *Archin*, Vol. 1, pp.20-21. In Macedonian.

Kovacev, P. (1987). *Urban Planning With Special Emphasis on the Earthquake as Natural Factor*, Unpublished graduate diploma dissertation, Universitet Kiril i Metodij: Skopje. In Macedonian.

Kovraz, M. (1985). 'Shelters in Urban Planning and Construction', in B. Vandot, (ed.), *Shelters Design: Application of Shelters Building Code*, Privredni Vjesnik: Zagreb, pp.1-34. In Serbo-Croatian.

Kreimer, A. (1988). *Guidelines for Banks Partricipation in Reconstruction Projects After Disasters*, World Bank: Washington, DC.

Krimgold, F. (1974). *Pre-disaster Planning: the Role of International Aid for Pre-disaster Planning in Developing Countries*, Avdelningen for Arlitektur KTH: Stockholm.

Krstic, B. (1987). *Urban Planning, Architectural, Conservation and Human Enironment Legislation*, Naucna Knjiga: Beograd. In Serbo-Croatian.

Kultermann, U. (1970). *Kenzo Tange - 1946-1969 - Architecture and Urban Design*, Pall Mall Press: London.

Lacey, L., and Owusu, S. (1988). 'Squatter Settlements in Monrovia, Liberia: The Evolution of Housing Policies', in R. A. Obudba and C. C. Mbanga, (eds.), *Slum and Squatter Settlements in Sub-Saharan Africa: Towards a Planning Strategy*, Prager: New York.

Ladinski, V.B. (1994a). 'Development Control Standards in Earthquake Prone Areas: A Case Study', in A. Awotona, (ed.), *Proceedings, An International Symposium on People, Place and Development*, Centre for Architectural Research and Development Overseas, Department of Architecture, University of Newcastle: Newcastle upon Tyne, pp.387-401.

Ladinski, V.B. (1994b). *Post 1963 Earthquake Housing Reconstruction in Skopje*, Unpublished Masters dissertation, Anglia Polytechnic University: Cambridge.

Ladinski, V.B. (1994c). 'Post 1963 Earthquake Reconstruction in Skopje', in A. Awotona, (ed.), *Proceedings, An International Symposium on People, Place and Development*, Centre for Architectural Research and Development Overseas, Department of Architecture, University of Newcastle: Newcastle upon Tyne, pp.372-386.

Ladinski, V.B. (1995a). 'Debris Management and Recycling Following the 1963 Skopje Earthquake: A Case Study', in *Recycling and Sustainable Post-Disaster Reconstructin Workshop*, Post-War Reconstruction and Development Unit, Institute of Advanced Architectural Studies, University of York: York.

Ladinski, V.B. (1995b). 'Development and Conservation Aspects of Urban Regeneration in Seismic Prone Areas: A Case Study.' *International Workshop on Urban Regeneration: Development Through Conservation: Towards Shaping World Cities*, Intitute of Advance Architectural Studies, University of York, UK, pp.167-181.

Lagorio, H.J. (1990). *Earthquakes: An Architect's Guide to Nonstructural Seismic Hazards*, John Wiley and Sons, Inc.: New York.

Laska, S.B. (1991). 'Floodproof Retrofitting: Homeowners Self-Protective Behaviour.' *Program on Environment and Behavior Monograph No.49*, Institute of Behavioral Science.

Lewis, J. (1981). 'Some perspectives on natural disaster vulnerability in Tonga.' *Pacific Viewpoint*, Vol. 22, No. 2. Victoria University of Wellington.

Lewis, J. (1984). 'Environmental interpretations of natural disaster vulnerability: The crucial need.' *The Environmentalist*, Vol. 4, No. 3, pp.177-180. Guest Editorial.

Lewis, J. (1987). 'Vulnerability and development - and the development of vulnerability; a case for management.' *Development Studies Association Conference*, University of Manchester.

Lewis, J. (1991). *The relief syndrome: post World War II and sequentialism*, Mimeo. Data International.

Lewis, J. (1994a). 'Disasters, conflict and recovery: What kind of development for sustainability and disaster reduction.' *Reconstruction after Disaster: The State of the Art*, CARDO: University of Newcastle (not presented).

Lewis, J. (1994b). 'Reconstruction, development and disaster reduction: Cyclone shelters in coastal Bangladesh.' *Reconstruction After Disaster: The State of the Art*, CARDO: University of Newcastle (not presented).

Lewis, O. (1976). *Five Families: Mexican Case Studies in the Culture of Poverty*, Souvenir Press: London.

Liberia. (1984). 'Rappateur's Report', in *Population and housing census data user's seminar*, MPEA: Monrovia, Liberia.

Liberia. (1993). *The Growing Problem of Displaced People in ECOMOG Controlled Areas. Annual Report of The National Repatriation and Resettlement Commission.* MPEA: Monrovia, Liberia.

Markov, P.J. (1982). 'Urban Planning and Earthquake Protection', in D. Zivkovic, (ed.), *Architectural Handbook 1982*, Savez Arhitekata Srbije: Beograd, pp.68-74. In Serbo-Croatian.

Markov, P.J. (1983). 'Revitalisation and Reconstruction of Buildings as Element of Protection in Urban Planning', in L. Kodelja, (ed.), *Rehabilitation and Reconstructin of Buildings*, Savez Arhitekata Jugoslavije: Beograd, pp.9-33. In Serbo-Croatian.

Maskrey, A. (1987). *Community Disaster Mitigation*, Oxfam: UK.

Mason, V. (1993). *A New Direction for Sustainable Shelter Development*, National Housing Authority: Monrovia, Liberia. Unpublished.

May, P.J., Burby, R.J., *et al.* (1996). *Environmental Management and Governance, International Approaches to Hazards and Sustainability*, Routledge: London.

McNeill, P. (1985). *Research Methods*, Tavistock: London.

Mihailov, V., and Talaganov, K. (1985). 'Seismic Hazard of Skopje Area.' *Edilizia Popolare*, Vol. 187, pp.11-17.

Milutinovic, Z., Trendafiloski, G., *et al.* (1995). *Disaster Preparedness Planning for Small and Medium Size Hospital Based on Structural, Nonstructural and Functional Vulnerability Assessment.* Report IZIIS 95-68, Institute of Earthquake Engineering and Engineering Seismology, University St. Cyril and Methodius, Skopje and World Health Organisation, Regional Office for Europe, Copenhagen: Skopje.

Mitchell, J.K. (1994). 'Natural disasters in the context of megacities.' *Megacities and Disasters*, United Nations University: Tokyo.

Montz, B., and Gruntfest, E.C. (1986). 'Changes in American urban floodplain occupancy since 1958: the experience of nine cities.' *Applied Geography*, Vol. 6, pp.325-338.

NARDA. (1992). *A report of the Round Table Conference on Strategies and Direction for the Reconstruction and Development of Liberia*, Unpublished: Monrovia, Liberia.

Norton, R. (1994). 'Macedonia Recovery Plan on World Table.' *European*, 196, p. 19.

Parker, D.J. (1992). *Flood proofing, Territorial Land Drainage and Flood Control Strategy Study Phase II.* Binnie and Maunsell Consultants, for Drainage Services Department, Hong Kong Government.

Parker, D.J. (1995). 'Floodplain development policy in England and Wales.' *Applied Geography*, Vol. 15, No. 4, pp.341-363.

Parker, D.J., and Fordham, M. (1996). 'An evaluation of flood forecasting, warning and response systems in the European Union.' *Water Resources Management*, Vol. 10.

Parker, D.J., Fordham, M., *et al.* (1994). 'Real-time Hazard Management: Flood Forecasting, Warning and Response', in E. C. Penning-Rowsell and M. Fordham, (eds.), *Floods Across Europe: Flood Hazard Assessment, Modelling and Management*, Middlesex University Press: London, pp.135-166.

Parker, D.J., Green, C.H., *et al.* (1987). *Urban flood protection benefits: a project appraisal guide*, Gower Technical Press: Aldershot, UK.

Parker, D.J., and Neal, J. (1990). 'Evaluating the Performance of Flood Warning Systems', in J. W. Handmer and E. C. Penning-Rowsell, (eds.), *Hazard and the Communication of Risk*, Gower Technical Press: Aldershot, UK, pp.137-156.

Parker, D.J., and Thompson, P.M. (1991). 'Floods and Tropical Storms', in *The Challenge of African Disasters*, World Health Organisation, Panafrican Centre for Emergency Preparedness and Response: Addis Ababa, Ethiopia, pp.38-59.

Patton, M.Q. (1990). *Qualitative Evaluation and Research Methods*, Sage: London.

Peattie, L. (1983). 'Realistic Planning and Qualitative Research.' *Habitat International*, Vol. 7, No. 5/6, pp.227-234.

Perrow, C. (1984). *Normal Accidents: Living with High-Risk Technologies*, Basic Books: New York.

Petrovski, J., and Milutinovic, Z. (1985). 'Earthquake vulnerability and risk assessment for physical and urban planning.' *Edilizia Popolare*, Vol. 187, pp.18-25.

Petrovski, J., and Milutinovic, Z. (1988). 'Vulnerability and Seismic Risk Assessment of Old Towns.' *1st International Seminar on Modern Principles in Conservation and Restoration of Urban and Rural Cultural Heritage in Seismic-Prone Regions*, Skopje, pp.153-164.

Plummer, K. (1983). *Documents of Life: an Introduction to the Problems and Literature of a Humanistic Method*, Allen & Unwin: London.

Polservice and Institute of Town Planning and Architecture (1965). *Skopje Urban Plan*, Vol. 15, Zavod za Urbanizam I Arhitektura: Skopje. In Macedonian.

Popovic, O. (1990). *Urban Complexes Renewal Planning Following Strong Earthquakes*, Unpublished Masters dissertation, Universitet Kiril i Metodij: Skopje. In Macedonian.

Quarantelli, E. (1978). *Disasters: Theory and Research.* Sage: London.

Rios, D. (1987). 'Barrios Altos Earthquake Study', in A. Maskrey, *Disaster Mitigation*, IT Publications: London.

Rogge, J.R. (1994). 'Repatriation of Refugees', in T. Allen and H. Morsink, (eds.), *When Refugees Go Home*, UNRISD: Geneva.

Rowland, J. (1995). 'Rebuilding Sarajevo', in *Disaster Prevention and Management: An International Journal*, Vol. 4, November 1.

SBA. (1996). *Monrovia Baseline Survey*, Vol. Unpublished: Monrovia, Liberia.

Schware, R. (1982). 'Official and Folk Flood Warning Systems: An Assessment.' *Environmental Managment*, Vol. 6, No. 3, pp.209-216.

Shaffir, W., and Stebbins, R. (1991). *Experiencing Fieldwork: An Inside View of Qualitative Research*, Sage: London.

Sheaffer, J.R. (1960). *Flood proofing: an element in a flood damage reduction program*. Department of Geography Research Paper 65, University of Chicago: Chicago.

Skartov, K., and Stojkov, T. (1985). 'Reconstruction and Construction of Skopje After the Catastrophic July 26, 1963 Earthquake.' *Edilizia Popolare*, Vol. 187, pp.5-10.

Skopje Faculty of Architecture. (1977). *Evaluation of Designed and Constructed Housing Estates*. Unpublished report Arhitektonsik Facultet: Skopje. In Macedonian.

Smith, K. (1992). *Environmental Hazards: Assessing Risk and Reducing Disaster*, Routledge: London.

Stacey, M. (1969). *Methods of Social Research*, Pergamon: Oxford.

Subah, P. (1994). *Manufacture of Fibre Concrete Roofing Tiles and Stabilized Soil Block: A Technical and Financial Assessment*, Subah Belleh Associates: Monrovia, Liberia. Unpublished.

Taylor, A. (1981). 'Assessment of Victim Needs', in I. Davis, (ed.), *Disasters and the Small Dwelling*, Pergamon: Oxford, pp.134-137.

Thompson, P.M., Wigg, A.H., *et al.* (1991). 'Urban flood protectin post-project appraisal in England and Wales.' *Project Appraisal*, Vol. 6, No. 2, pp.84-92.

Tolba, M. (1992). *Saving Our Planet: Challenges and Hopes*, Chapman & Hall: London.

Turner, B.A. (1978). *Man-Made Disasters*, Wykeham: London.

UNCED. (1993). *Report of the United Nations Conference on Environment and Development.* , Rio de Janiero 1992.

UNHCR. (1995). *The State of Sierra Leonean Refugees and Displaced Liberians.* Unpublished UNHCR: Monrovia, Liberia.

UNCHS. (1987). *Global Report on Human Settlements 1986*, Oxford University Press: Oxford.

UNCHS. (1994). 'Sustainable Human Settlements in an Urabnising World, Including Issues Related to Land Policies and Mitigatrion of Natural Disasters.' *15th Session of the Commission on Human Settlements.* Unpublished draft theme paper.

UNDRO. (1982a). *Disaster prevention and mitigation: a compendium of current knowledge*, Vol. 8, UN: New York.

UNDRO. (1982b). *Shelter After Disaster: Guidelines for Assistance*, UNDRO: New York.

UNDRO. (1991a). 'Editorial.' *UNDRO News*, Sept/Oct.

UNDRO. (1991b). *Mitigating Natural Disaster: Phenomena, Effects and Options. A Manual for Policy Makers and Planners*, UNCHS: Geneva.

Verney, P. (1979). *The Earthquake Handbook*, Paddington Press Ltd.: New York.

Vesilind, P.J. (1996). 'Macedonia: Caught in the Middle.' *National Geographic*, Vol. 189, No. 3, pp.118-139.

Vlcevski, R. (1984). *Urbo-Sociological Aspects of Residential Settlements Planning in Skopje*, Unpublished Masters dissertation, Universitet Kiril i Metodij: Skopje. In Macedonian.

Walker, P. (1995). 'Defending Humanitatianism.' *The Courier*, Vol. 150, March-April 1995.

Walker, R. (1985). *Applied Qualitative Research.* Gower: Aldershot.

Westcoat, J.L., Chowdhury, J.U., *et al.* (1992). 'Five Comments on the Bangladesh Flood Action Plan.' *Natural Hazards Observer*, Vol. 16, No. 4, pp.1-7.

White, G.F. (1945). *Human adjustment to floods*. Research Paper 29, Department of Geography, University of Chicago: Chicago.

White, G.F., Calef, W.C., *et al.* (1958). *Changes in Urban Occupance of Flood Plains in the United States.* Research Paper 57, Department of Geography, University of Chicago: Chicago.

Whyte, W.F. (1984). *Learning From the Field: a Guide From Experience*, Sage: London.

Wijkman, A., and Timberlake, L. (1984). *Natural Disasters: Acts of God or Acts of Man?*, Earthscan: London.

Winchester, P. (1992a). *Power, Choice and Vulnerability: A Case Study in Disaster Mismanagement in South India*, James and James: London.

Winchester, P. (1992b). *Power, choice and vulnerability: a case study of disaster managment*, James and James: London.

Wisner, B. (1995). 'Bridging "Expert" and "Local" knowledge for counter-disaster planning in urban South Africa.' *Geojournal*, Vol. 37, No. 3, pp.335-348.

World Bank. (1989). *Reaching the Poor through Urban Operations*, World Bank - Infrastructure and Urban Development Department: Washington, DC.

World Commission on Environment and Development. (1987). *Our Common Future*, Oxford University Press: Oxford.

Yin, R. (1989). *Case Study Research: Design and Methods*, Sage: London.

Yong, C., Tsoi, K., *et al.* (1988). *The Great Tangshan Earthquake of 1976: An Anatomy of Disaster*, Pergamon Press: Oxford.

Zakic, N. (1982). 'Shelter as an Element of Urban Planning and Design', in D. Zivkovic, (ed.), *Architectural Handbook 1982*, Savez Arhitekata Srbije: Beograd, pp.128-144. In Macedonian.

Zakic, N. (1983). 'Protection and Recovery from Collapsed Buildings', in L. Kodelja, (ed.), *Rehabilitation and Reconstruction of Buildings*, Savez Arhitekata Jugoslavije: Beograd, pp.35-47. In Serbo-Croatian.

Zargar, A. (1989). 'War as a Disaster', in C. Cockburn and A. Zargar, (eds.), *Settlements Reconstruction After War*, University of York: York.

# Index